Science Buddies
Cooperative Science Activities

Laura Candler

Kagan

1996, 2000 by *Kagan Publishing*

This book is published by *Kagan Publishing*. All rights are reserved by *Kagan Publishing*. No part of this publication may be reproduced or transmitted in any form by any means, electronic or mechanical, including photocopy, recording, or any information storage and retrieval system, without prior written permission from *Kagan Publishing*. The blackline masters included in this book are intended for duplication only by classroom teachers who purchase the book, for use limited to their own classrooms. To obtain additional copies of this book, or information regarding other *Kagan* products or professional development, contact:

Kagan Publishing
981 Calle Amanecer
San Clemente, CA 92673
1 (800) 933-2667
www.KaganOnline.com

ISBN: 978-1-879097-37-7

Table of Contents

- *Acknowledgments* .... *II*
- *Introduction* .... *III*
- *Foreword* .... *IV*

CHAPTER 1
The Basics

- Steps to Implement the Program 2
- Strategies for Success 3
- Science Buddies Program Planner 4
- Science Buddies Topic Reference 5
- Science Buddies Process Skills 6
- Record-Keeping and Assessment 7
- Creating a Cooperative Classroom 8

CHAPTER 2
Cooperative Learning Structures

- Science Buddies Structure Reference 10
- Blackboard Share 11
- Mix-Freeze-Pair 12
- Numbered Heads Together 13
- Pair Discussion 14
- RoundRobin 15
- Team Discussion 16
- Think-Pair-Share 17

CHAPTER 3
Activities

- Chilly Beans 20
- Yeast Power 22
- Worm Hunt 24
- Egg-citing Egg Trick 26
- Tricky Twirler 28
- Taste Test 30
- Hot and Cold Race 32
- Bubble Mania 34
- Crystal Creations 36
- Ice Cream Investigations 38
- Silly Slime 40
- Cabbage Juice Chemistry 42
- Raisin Razzmatazz 44
- Creeping Colors 46
- The Mysterious Balloon 48
- Sun Fun 50
- Color Spinners 52
- The Magic Card 54
- The Amazing Paper Kite 56
- Wonderful Wind Socks 58
- Weathering and Erosion Walk 60
- Balloon Blast-Off 62
- Straw Oboes 64
- Lever Logic 66

CHAPTER 4
Additional Reproducibles

- Letter to Parents 70
- Hints for Science Buddies 71
- Science Buddies Lab Report (Full Page) 72
- Science Buddies Lab Report (Half Page) 73
- Science Buddies Roster 74
- Super Science Buddies Award 75
- Color Spinner Patterns 76
- The Amazing Paper Kite Pattern 77
- Weathering and Erosion Log 78

Science Buddies® Laura Candler • Kagan Publishing • 1-800-933-2667

Foreword

Dr. Spencer Kagan

Great gifts sometimes come in small packages. Science Buddies is one such gift.

This gift goes to teachers, students, parents, and a nation urgently in need of reforming its science curriculum — as well as the way parents and children spend time together.

The idea is simple: Send home simple science experiments which a student conducts and discusses with a "Science Buddy" — a parent or older, caring adult.

At a time when the nation's youths spend more hours watching television than any other single waking activity (1000 classroom hours a year, but nearly 1500 TV hours), Science Buddies gets a child and a parent to take a bit of time out to explore and think together. Is it possible to send home simple homework assignments that transform both attitudes toward science and child-parent relationships? Can homework be a tool for restructuring entrenched family interaction patterns so in need of reform?

Today, parents and children sit for hours, side by side, focused on the television. With regard to communication, they might as well be in different rooms (or on different planets). After a television program, what do families do? Watch another. And another — until they are tired enough to go to bed. Perform the following experiment: After the first television program of the evening, walk over to the set and turn it off. Turn to the family members and say, "Let's discuss the program." The strange expressions on their faces say it all: Communication among family members is no longer the norm. Our nation's families have lost the art of conversation — talking, thinking together is now preempted by violent and meaningless programming.

Viewed in this light, Science Buddies is a daring experiment.

And the results are powerful: Students and their Science Buddy learn a range of science concepts and skills, open the doors to questioning and thinking, spend quality time together, and make the science curriculum more approachable. Laura Candler's simple send-home science projects transform the science curriculum. The metacommunication to students: Science is not a forbidding, mysterious activity conducted by unapproachable scientists in distant labs with complicated apparatus; it is a fun, approachable process — conducted with everyday materials, by any of us, any time, with a buddy.

Project 2061 is a long term, multiphase undertaking of the American Association for the Advancement of Science. The goal: create science literacy for all citizens. "A cascade of recent studies has made it abundantly clear that by both national standards and world norms, U. S. education is failing to adequately educate too many students — and hence failing the nation." In response, the leading science educators of the nation call for comprehensive curriculum reform. They note reform must be collaborative; parent involvement is a critical element. They ask that science curricula be transformed, breaking down rigid subject-matter boundaries, de-emphasizing discrete details, emphasizing instead main concepts and presenting "the scientific endeavor as a social enterprise."

Science Buddies is a powerful tool for revitalizing our science curriculum ... and the way students and parents spend their together time.

<u>Reference</u>
American Association for the Advancement of Science. Science for All Americans. A Project 2061 Report on Literacy Goals in Science, Mathematics, and Technology. American Association for the Advancement of Science. Washington, D.C., 1989.

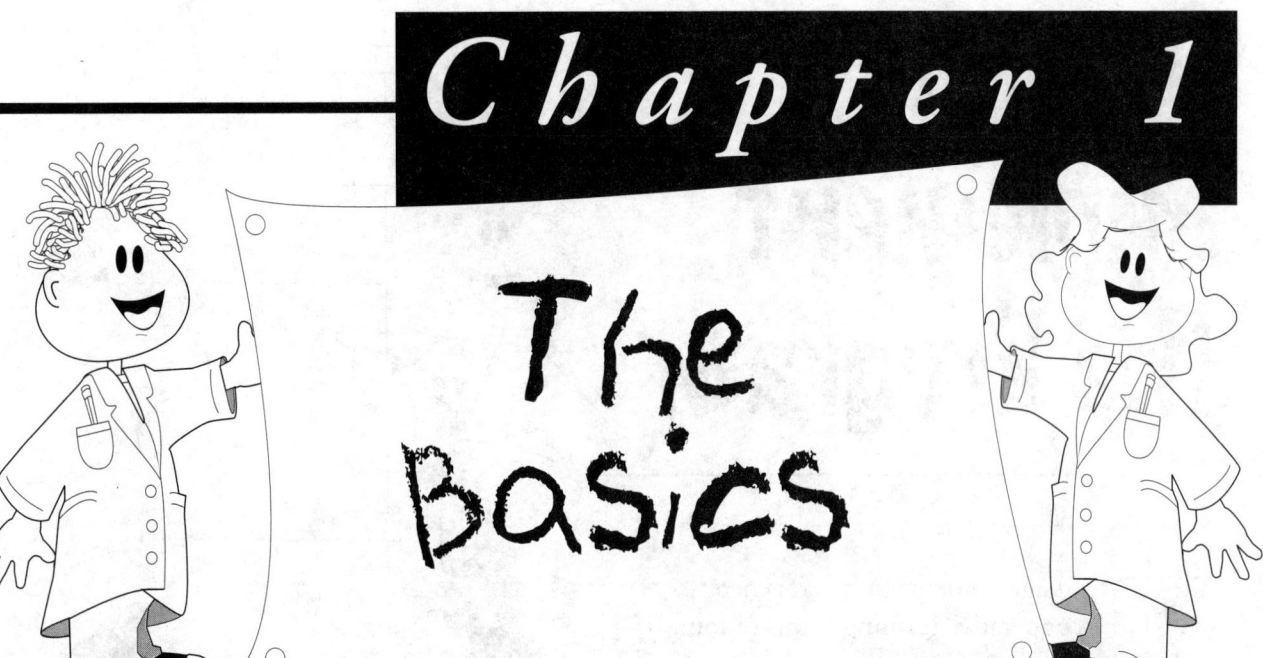

Chapter 1: The Basics

This chapter includes everything you need to know to implement the Science Buddies Program. You'll find ways to adapt the program to meet your curriculum goals, useful planning guides, and tips for record-keeping. In addition, this chapter provides basic information on setting up a classroom for cooperative learning.

Steps to Implement the Program

 Assign students to heterogeneous cooperative learning teams of four. Refer to "Creating a Cooperative Classroom" on page 8 for more information on team formation and cooperative classroom management.

 Select your first activity and decide on a due date. Two weeks is generally sufficient. Write the due date on the activity page before you duplicate copies for your students.

 Choose a method for students to report results. Consider one of the Lab Report forms, a science journal entry, or an illustrated paragraph.

 Introduce the activity to students by using the cooperative techniques described on the Teacher's Page. Refer to the Cooperative Learning Structure descriptions on pages 11-17 for more detailed information about each cooperative technique.

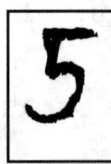 Introduce the program to parents by sending home the Parent Letter and "Hints for Science Buddies" along with your first activity (see pages 70-71).

 On or after the due date, use the cooperative classroom follow-up strategies to discuss and extend the concepts. This component of the program ensures that students receive accurate information about their science explorations. Allow time for sharing and discussion between students.

 Send home a new Science Buddies activity on the first school day of each month.

 At the end of the year, award Super Science Buddy certificates (page 75) to students who have completed all activities.

Strategies for Success

 Keep the program simple and fun by limiting the amount of written work assigned along with it. Make sure students understand that the "Talk it Over" questions at the end of each activity are for discussion only. Discourage lengthy, written responses or students may begin to view the program as just another dreaded homework assignment.

 Plan a year's worth of Science Buddies activities in advance, using the Science Buddies Program Planner. Consider factors such as your curriculum goals and the season of the year.

Refer to the Process Skills checklist on page 6 to find the science skills used in each Science Buddies activity. Choose activities that reinforce the process skills you are using in class, such as observing, measuring, organizing data, or experimenting.

 Don't allow too much time for completion of activities at home. Two weeks is usually enough. Experience has shown that when given more time, students tend to lose their activity page or forget to complete the assignment.

 If students seem to have trouble obtaining materials, ask your local parent-teacher organization to sponsor your Science Buddies Program. Materials like straws, dried beans, balloons, and index cards can be cheaply purchased in class-sized quantities, sparing parents the time and expense of hunting for such items.

 Try to relate each activity to everyday situations that occur at home or in the classroom. Science becomes more meaningful when students observe its effects on their lives.

Science Buddies® Laura Candler • Kagan Publishing • 1-800-933-2667

Science Buddies Program Planner

Use this worksheet to plan a year's worth of Science Buddies activities. Review your curriculum goals for each month and select activities which will enrich your science program.

Month   **Activity**

September _____

October _____

November _____

December _____

January _____

February _____

March _____

April _____

May _____

June _____

July _____

August _____

4   Science Buddies® Laura Candler • Kagan Publishing • 1-800-933-2667

Science Buddies
Topic Reference

Activities | Topics

Activity	Topic
Chilly Beans	Seeds and Plants
Yeast Power	Living Things
Worm Hunt	Animal Adaptations
Egg-citing Egg Trick	Animals
Tricky Twirler	Human Body
Taste Test	Human Body
Hot and Cold Race	Molecules
Bubble Mania	Surface Tension
Crystal Creations	Solutions and Crystals
Ice Cream Investigations	Matter
Silly Slime	Matter
Cabbage Juice Chemistry	Matter and Chemistry
Raisin Razzmatazz	Chemistry
Creeping Colors	Chemistry
The Mysterious Balloon	Static Electricity
Sun Fun	Solar Energy
Color Spinners	Light and Color
The Magic Card	Air Pressure
The Amazing Paper Kite	Air and Flight
Wonderful Wind Socks	Air and Weather
Weathering and Erosion Walk	Soil Erosion
Balloon Blast-Off	Space
Straw Oboes	Sound
Lever Logic	Simple Machines

Science Buddies Process Skills

Activities	Observing	Predicting	Making Models	Measuring	Organizing Data	Inferring	Communicating	Experimenting
Chilly Beans	•	•		•	•	•	•	•
Yeast Power	•	•		•		•	•	
Worm Hunt	•			•	•	•		
Egg-citing Egg Trick	•	•				•	•	
Tricky Twirler	•		•		•		•	
Taste Test	•				•		•	
Hot and Cold Race	•	•	•	•		•	•	•
Bubble Mania	•			•		•	•	
Crystal Creations	•			•		•	•	
Ice Cream Investigations	•	•		•		•	•	
Silly Slime	•					•	•	•
Cabbage Juice Chemistry	•			•	•	•	•	
Raisin Razzmatazz	•	•				•	•	
Creeping Colors	•					•	•	•
The Mysterious Balloon	•	•			•	•	•	
Sun Fun	•	•	•	•	•	•	•	•
Color Spinners	•	•		•		•	•	•
The Magic Card	•	•				•	•	
The Amazing Paper Kite	•		•	•		•	•	
Wonderful Wind Socks	•		•	•			•	
Weathering and Erosion Walk	•				•	•	•	
Balloon Blast-Off	•	•	•			•	•	
Straw Oboes	•	•		•		•	•	•
Lever Logic	•	•	•	•	•	•	•	•

6 Science Buddies® Laura Candler • Kagan Publishing • 1-800-933-2667

Record-Keeping and Assessment

 Remember that the Science Buddies program is designed to foster positive attitudes by allowing students and parents to explore science together. Don't spoil the fun by requiring lengthy written reports!

 Decide whether the activities will be required or optional. Offering extra credit for optional assignments may encourage participation.

 Use the Science Buddies Roster to keep a list of students who have completed each activity. At the end of the year, present a Super Science Buddy Award to each student who has completed all activities.

 Choose a method for having students report the results of their activities. Two Lab Report forms are provided, or you may design your own. If students have science journals, consider having them write a brief description of their results in their journals.

 If you grade Lab Reports or journal entries, don't penalize students for activities that didn't work. Ask them to describe what did happen and write why they think the activity didn't work. (Perhaps they didn't follow directions or substituted materials that changed the outcome.)

 Be sure to schedule time for discussing each activity on its due date. This is a great time for informally assessing student understanding of critical science concepts. Clarify any misunderstandings and encourage students to share what they learned with their Science Buddies at home.

Science Buddies® Laura Candler • Kagan Publishing • 1-800-933-2667

7

Creating a Cooperative Classroom

Team Formation

- Divide your class into teams of four students. Four is the optimal number since a team of four can easily be divided into two sets of pairs.

- Make sure teams are as heterogeneous as possible. Consider factors such as academic ability, race, gender, and personality.

- Seat team members together by clustering desks or using tables.

- Form new teams at least every 4 to 6 weeks.

Management Ideas

- Implement a Quiet Signal to use when you need your students' attention. Raising your hand, clapping a pattern, and ringing a bell are all effective when used consistently.

- Monitor team interactions and encourage equal participation. If several teams are having difficulty, trying dividing the class into partners for pair activities. After students master the social skills needed for pair work, form new teams of four and try again.

- For easy distribution, keep the materials for each team in an accessible location. Try placing a basket or margarine tub with glue, markers, crayons, scrap paper, and scissors in the center of each team.

- With your students, establish rules for your cooperative classroom and expect students to abide by those rules. Post the rules clearly.

Our Cooperative Classroom Rules

1. Listen to each other.
2. Obey the Quiet Signal.
3. Take turns.
4. Share materials.
5. Praise each other. (No put-downs allowed!)

Chapter 2

Cooperative Learning Structures

Cooperative Learning Structures are teaching techniques that can be adapted to almost any subject matter. The seven structures described here are primarily used for fostering discussion, both before each Science Buddies activity is distributed, and after it has been completed at home.

Science Buddies Structure Reference

Activities	Blackboard Share	Mix-Freeze-Pair	Numbered Heads	Pair Discussion	RoundRobin	Team Discussion	Think-Pair-Share
Chilly Beans	•					•	•
Yeast Power			•		•	•	
Worm Hunt		•					•
Egg-citing Egg Trick		•			•		
Tricky Twirler		•	•				•
Taste Test	•				•		
Hot and Cold Race		•				•	
Bubble Mania			•	•	•		
Crystal Creations						•	•
Ice Cream Investigations		•		•			
Silly Slime						•	•
Cabbage Juice Chemistry					•	•	•
Raisin Razzmatazz	•		•	•			
Creeping Colors	•		•		•	•	
The Mysterious Balloon				•		•	
Sun Fun			•	•			
Color Spinners		•				•	•
The Magic Card			•	•			
The Amazing Paper Kite				•	•		•
Wonderful Wind Socks		•			•		
Weathering and Erosion Walk				•	•		•
Balloon Blast-Off							
Straw Oboes		•					
Lever Logic			•	•			

10   Science Buddies® Laura Candler • Kagan Publishing • 1-800-933-2667

Cooperative Learning Structure
Blackboard Share

Blackboard Share is an excellent structure for reporting investigation results. Students prepare team charts or graphs of their data on newsprint or a section of the chalkboard. Team members share responsibilities or the teacher assigns roles such as:

• Title Writer	Writes title in bold letters
• Chart Maker	Draws blank chart or graph
• Data Recorder	Fills in chart with results
• Illustrator	Adds border or pictures
• Reporter	Presents results orally to class

Steps

1 Teacher announces topic and method of sharing.

2 Teams prepare for sharing session.

3 Teams share responses with the class.

Science Buddies® Laura Candler • Kagan Publishing • 1-800-933-2667

Cooperative Learning Structure
Mix-Freeze-Pair

This structure offers additional benefits beyond its value as a tool for discussion. By repeatedly mixing around the room and forming partners, students are interacting with classmates other than their own team members. **Mix-Freeze-Pair** is also a great energizer to use when students have been sitting or concentrating for long periods of time.

Steps

1. Teacher announces "Mix!" and students mill around the classroom.
2. Teacher calls "Freeze!" and students stop.
3. Teacher announces "Pair!" and students find a partner.
4. Teacher announces discussion topic or task for pair work.
5. Pairs discuss topic or perform task.
6. Students mix, freeze, and pair for each new topic or task.

Cooperative Learning Structure
Numbered Heads Together

Numbered Heads Together is an excellent structure for answering science questions. After a number is called, designated students may respond by writing on individual slates, giving oral answers, or holding up prepared response cards. To ensure individual accountability, students may not receive help from their teammates after a response number is chosen.

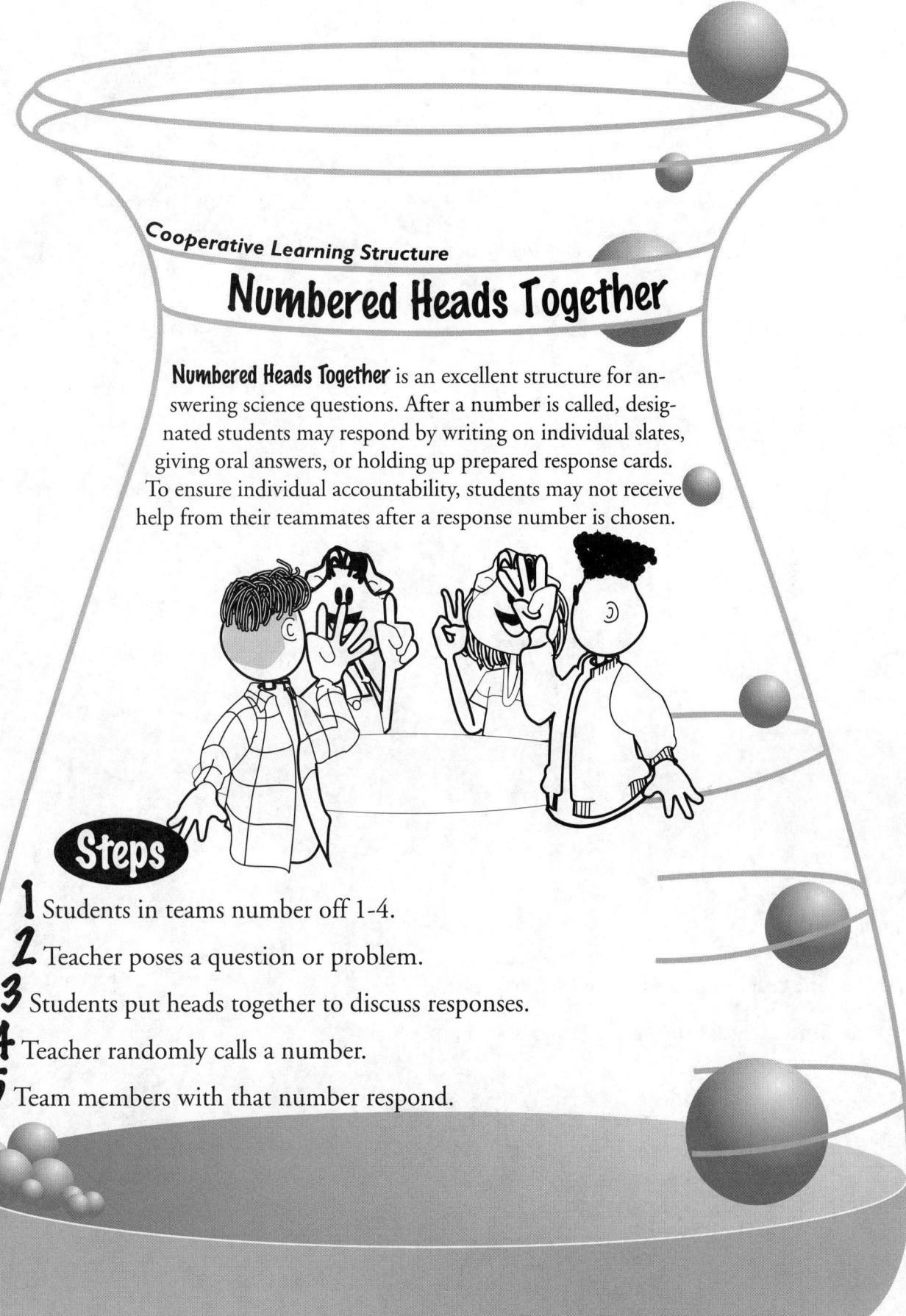

Steps

1. Students in teams number off 1-4.
2. Teacher poses a question or problem.
3. Students put heads together to discuss responses.
4. Teacher randomly calls a number.
5. Team members with that number respond.

Science Buddies® Laura Candler • Kagan Publishing • 1-800-933-2667

Cooperative Learning Structure

Pair Discussion

Pair Discussion is a quick and easy structure, used when informal discussion between students is the primary goal. If used during the direct instruction of science concepts, any Pair Discussion should be followed with a Class Discussion to clarify misconceptions.

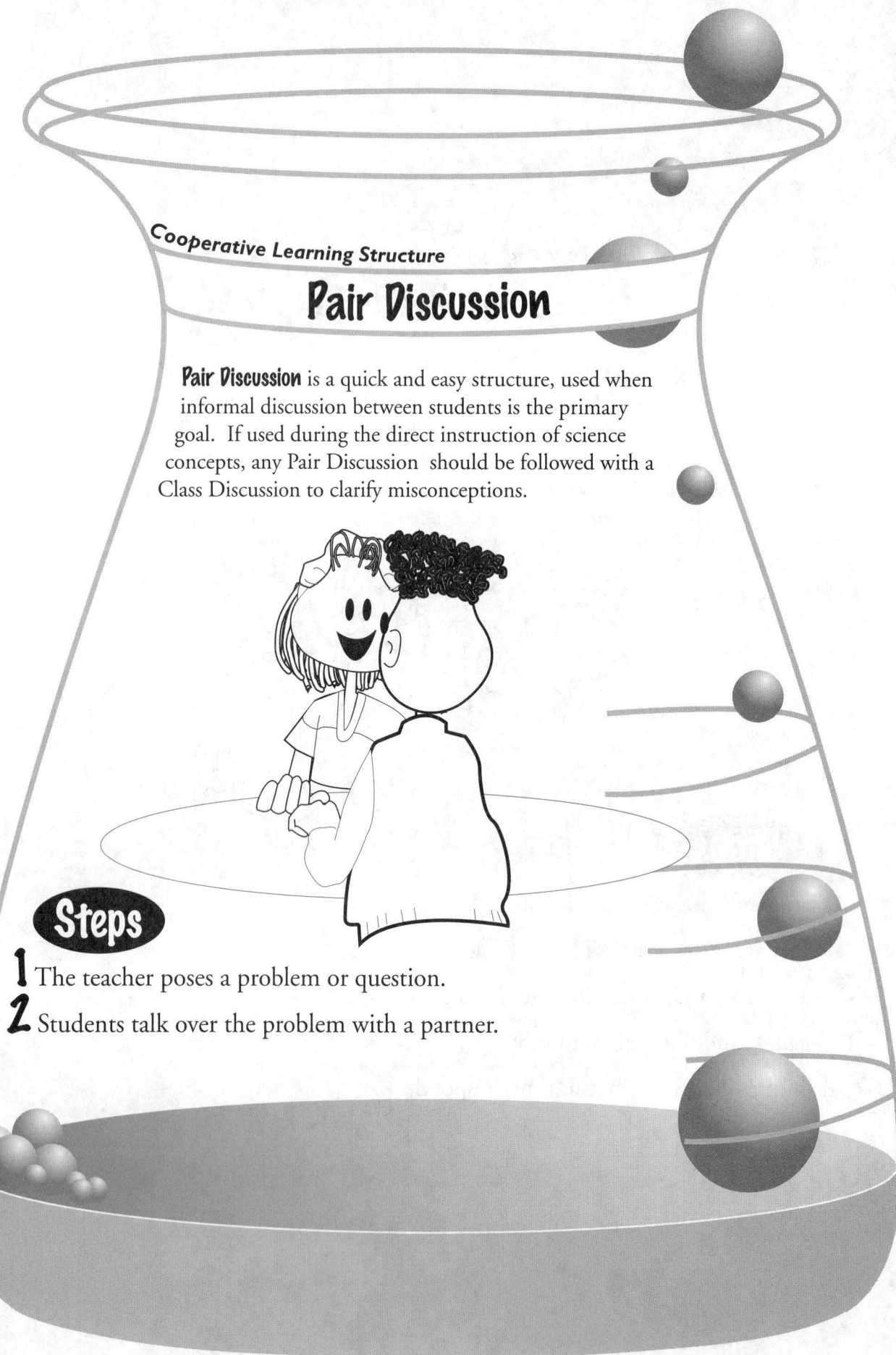

Steps

1. The teacher poses a problem or question.
2. Students talk over the problem with a partner.

Cooperative Learning Structure
RoundRobin

RoundRobin is a simple structure that encourages equal participation when students are responding to a question or prompt. Often students are given a specific amount of time to respond, such as 30 seconds or a minute. RoundRobin is not a discussion, however. Students focus on actively listening to their peers, rather than debating or discussing their responses.

Steps

1. Students number off 1 - 4 within teams.
2. Teacher announces topic or question.
3. Students take turns sharing their responses in numerical order.

Science Buddies® Laura Candler • Kagan Publishing • 1-800-933-2667

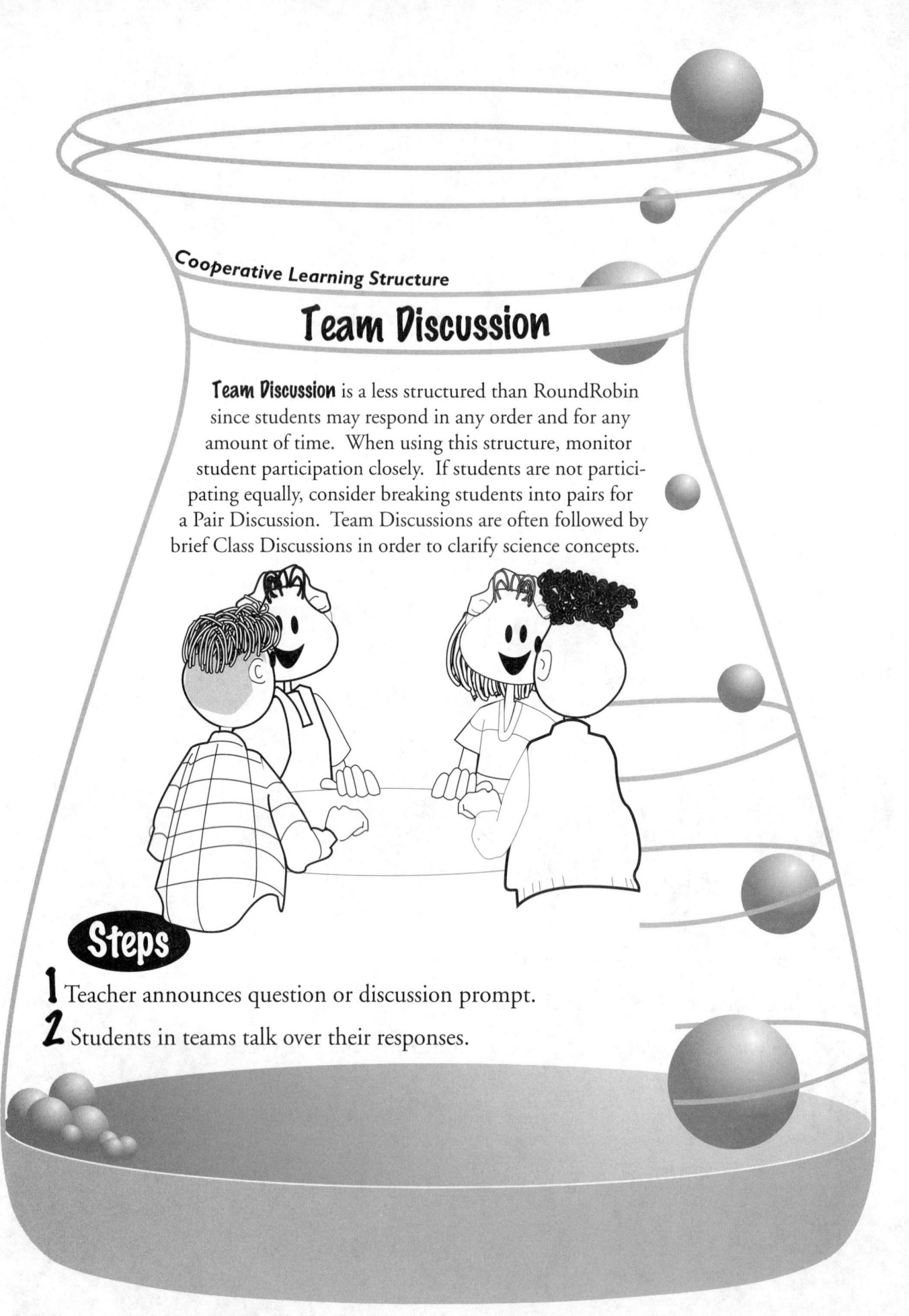

Cooperative Learning Structure

Team Discussion

Team Discussion is a less structured than RoundRobin since students may respond in any order and for any amount of time. When using this structure, monitor student participation closely. If students are not participating equally, consider breaking students into pairs for a Pair Discussion. Team Discussions are often followed by brief Class Discussions in order to clarify science concepts.

Steps

1. Teacher announces question or discussion prompt.
2. Students in teams talk over their responses.

Cooperative Learning Structure
Think-Pair-Share

Think-Pair-Share is an ideal structure for science instruction. Giving think time encourages higher-level processing and increases the quality of responses. The pair discussion allows students to become aware of new perspectives. During the class "sharing" phase, teachers can identify and correct science misunderstandings or challenge students to broaden their thinking.

1. The teacher poses a problem or question.
2. Students are given 5-10 seconds in which to think of individual responses.
3. Students pair with an assigned partner to discuss responses.
4. The teacher calls on several students to share their answers with the class.

Chapter 3

Activities

These 24 activities are designed to be completed at home with the help of an adult Science Buddy. Each activity has a teacher's page and a reproducible activity page. The teacher's page includes an overview, ideas for introducing the activity, cooperative classroom follow-up, and answers to questions on the activity page. The activity page lists the materials and the procedures for doing the activity.

Science Buddies Teacher's Page

Chilly Beans

Overview

Children are always fascinated when they sprout seeds and observe the young plants' growth. In this activity students experiment and discover that cold temperatures keeps seeds from sprouting. Students use the results of their experiments to solve the mystery of why seeds sprout in the springtime rather than during winter.

Introducing the Activity

Before handing out the Science Buddies activity, assign partners and use **Think-Pair-Share** to discuss the questions below. Instead of giving the answer to the second question, tell them that they will discover at least one answer during their Science Buddies activity.

- *What time of year do seeds sprout and flowers begin to bloom?*
- *Why don't plants bloom and grow in the winter?*

Cooperative Classroom Follow Up

Use **Blackboard Share** to discuss the charts students made in Step 8 of the activity (see sample). Give each team a large sheet of paper or a section of the chalkboard. Have them discuss the best type of chart to display their results. Then choose one person on each team to serve as the Recorder and to draw the chart, using their own data. Then select a team member to serve as the Reporter and present the chart to the class. Use **Team Discussion** to talk over the answers to the activity questions.

Sample Data

Day	Cold	Warm
1	0	8
2	0	10
3	0	10

Answers To Talk It Over

1. Predictions will vary, but results should be the same. The beans in the warm environment will sprout, but those in the refrigerator will not germinate.
2. If you switch the cups, the beans that were in the refrigerator will begin to sprout after being placed in the cabinet. Not surprisingly, the sprouted beans will stop growing when placed in the refrigerator.
3. Many factors affect the growth of seeds and plants in the spring. As shown in the Science Buddies activity, temperature change is an important factor. Longer day length is another factor, as well as an increase in moisture due to melting snow and increased rainfall.

Science Buddies® Laura Candler • Kagan Publishing • 1-800-933-2667

Science Buddies Activity

Chilly Beans

Procedures

1. Soak you bean seeds in water for several hours before starting the activity.
2. Since you will need to prepare two seed-sprouting cups for this experiment, you and your Science Buddy should *each* make one seed cup.
3. For each seed cup, fold a paper towel in thirds and then roll it into a tube. Slide it down the inside of the cup so that it lines the inside. Crunch up the other paper towels and stuff one into the middle of each cup.
4. Pour 1/3 cup water into each cup. Wait several minutes for the water to dampen the paper towels.
5. Place 10 seeds around the inside of each cup between the side and the paper towel liner. The seeds must *not* be below the water level.
6. Put one cup in the refrigerator and one cup in a warm dark cabinet.
7. Predict what will happen in each cup. Explain your prediction.
8. Observe the cups each day for 3 days. If the paper towel dries out, add a small amount of water. Make a chart showing the total number of seeds that have sprouted in each cup each day.

Due Date

Materials

- 20 dried bean seeds (limas, pintos, black beans, etc.)
- 2 clear cups or glasses
- 4 paper towels
- 2/3 cup water
- Measuring cup

 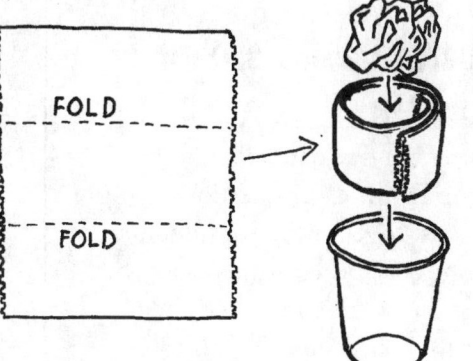 

Talk It Over

1. What did you predict would happen? What *did* happen?
2. What would happen if you switched the location of the two cups? Try it and see! Make your observations for several more days.
3. Why do you think seeds begin to sprout and grow in the springtime rather than in the winter?

Science Buddies® Laura Candler • Kagan Publishing • 1-800-933-2667

21

Science Buddies Teacher's Page

Yeast Power

Overview

This activity introduces students to a common microorganism, yeast, which is a member of the fungus family. After adding warm water to dry yeast and sugar, students watch a balloon inflate from the carbon dioxide gas produced.

Introducing the Activity

Before handing out the Science Buddies activity, give a slice of bread to each team and ask them to pass it around while they **RoundRobin** their observations. Encourage them to continue passing the bread until they are unable to state any more observations. Then pose these questions for a **Team Discussion**:
- *What ingredients do you think are used in making bread?*
- *What do you think made the holes you observed in the slice of bread?*

Tell students that they will learn more about bread-making as they complete their Science Buddies activity. Since the water temperature is very important in this activity, a thermometer is recommended. You might consider letting students check out classroom thermometers if your school has them.

Safety Notes: Caution students to be very careful when measuring the hot water.

Cooperative Classroom Follow Up

On or after the due date, discuss the answers to the Talk It Over questions using **Numbered Heads Together**. Have students number off from 1-4 in their teams. Pose the first question and ask students to put their heads together to discuss their responses. When all teams are ready, announce a number. The student on each team with that number stands to respond. Call on one person to start the answer and let each person standing add to that response. Repeat the sequence with the remaining three questions.

Answers To Talk It Over

1. Answers will vary.
2. Yeast is a tiny organism and a member of the fungus family. Unlike plants, yeast cannot make its own food. Instead, it breaks down sugars in a process called fermentation. Carbon dioxide gas is given off during fermentation.
3. Sugar is very important — yeast will not begin the fermentation process without sugar. An experiment can easily be designed to test the importance of sugar.
4. Yeast is responsible for making the bread light and airy. The carbon dioxide gas given off by the yeast during fermentation makes the bread rise.

Science Buddies Activity

Yeast Power

Procedures

1. Pour the sugar and dry yeast into the soda bottle.
2. Stretch the balloon by blowing it up and then letting all the air out. This will help it expand later.
3. Measure 1 cup very warm water. If you have a thermometer, adjust the water temperature so that it is between 105° and 115° Fahrenheit (40°- 50° Celsius)
4. Pour the water into the bottle and quickly fasten the mouth of the balloon over the opening of the bottle.
5. Predict what will happen to the yeast mixture.
6. Observe closely for 15 minutes. Write a list of your observations. (If nothing happens, start over with new materials and try hotter water.)
7. After 15 minutes, carefully remove the balloon. Smell the yeast mixture. What does it smell like?

Due Date

Materials

- 1 clean empty soda bottle (10-20 ounces in size)
- 1 balloon
- 1 package dry yeast
- 1 cup very warm water
- 1 tablespoon sugar
- thermometer (optional)
- clock or timer

Talk It Over

1. What observations did you make during this activity?
2. Why do you think the yeast mixture behaved as it did?
3. How important is the sugar in this activity? Could you design an experiment to find out?
4. Yeast is used in making bread. What job do you think yeast has in the bread-making process?

Science Buddies® Laura Candler • Kagan Publishing • 1-800-933-2667

23

Science Buddies Teacher's Page

Worm Hunt

Overview

Students discover the importance of camouflage by hunting for green and white "worms" on grass. This activity makes an effective introduction to the study of animal adaptations. Be sure to plan it for late spring, summer, or early fall when the grass is a healthy green color.

Introducing the Activity

Before handing out the Science Buddies activity, divide each team into two sets of partners and use **Think-Pair-Share** to discuss this question: *"What are some ways that animals escape their predators?"* Distribute the activity page and tell students that they will discover one way that some animals survive in their environments.

Cooperative Classroom Follow Up

Ask students to bring the chart of their results to class on the due date. Use **Mix-Freeze-Pair** to discuss the Talk It Over questions. Let students take their charts with them when you announce "Mix!" After allowing time for students to mingle, say "Freeze!" Then tell students to pair with a classmate and compare charts. Have them discuss the answer to the first question only. Repeat the **Mix-Freeze-Pair** sequence for each question.

Answers To Talk It Over

1. In most cases, students will collect more white "worms" than green ones. This is because the green worms blend in with the grass and are not as easy to see.
2. Results will be different depending on the surface color of the hunting ground.
3. Animals that blend in with their surroundings are not easily seen by their predators. This adaptation is known as "camouflage."
4. Answers will vary, but may include the following:
   • Speed
   • Sharp teeth or claws
   • Ability to fly or swim
   • Armor-like shell or sharp quills
   • Mimicry (looking like something in the environment such as a leaf or twig)

Science Buddies Activity

Worm Hunt

Procedures

1. Make 10 white and 10 green "worms" by cutting the yarn into 4-inch pieces. (If you don't have yarn, cut 10 white and 10 green worms out of construction paper.)
2. Measure a piece of white yarn 30 feet in length. Find an area with green grass and place the yarn in a circle on the grass. This will be your hunting ground.
3. Without letting your Science Buddy see, scatter all 20 worms within the yarn circle.
4. Give your Science Buddy 15 seconds to pick up as many worms as possible.
5. Count the number of worms *of each color* that your Science Buddy found. Record those numbers.
6. Now have your Science Buddy scatter all 20 worms on the grass for you to find. Hunt for 15 seconds, then record the number of worms you found of each color. Make a chart of your results.

Due Date

Materials
- Green and white yarn or construction paper
- Scissors
- Ruler or yardstick
- Timer or watch with second hand
- Pencil and paper
- Grassy area

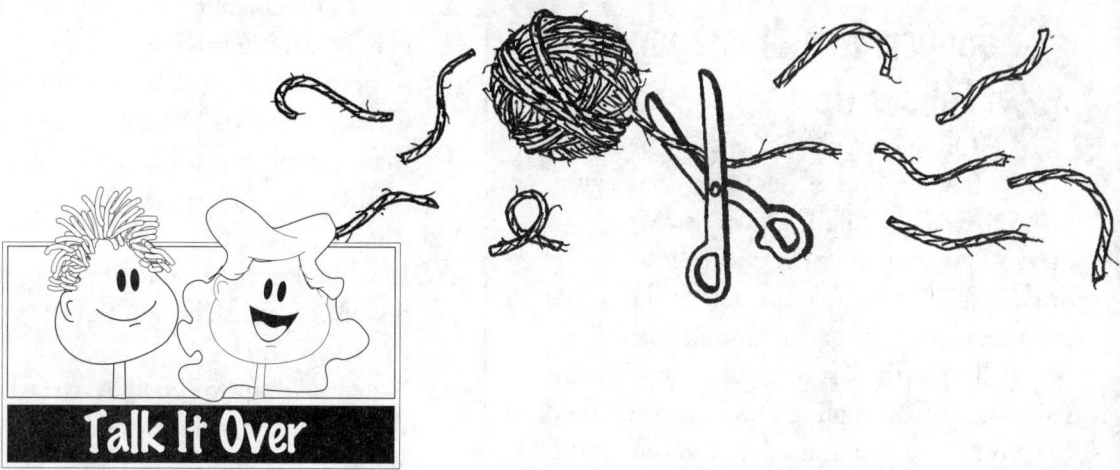

Talk It Over

1. How many white worms did you and your Science Buddy find? How many green worms? How can you explain your results?
2. Do you think you would get the same results if you moved your hunting ground to an area of dirt or pavement? Try it and see!
3. How can an animal's color help it survive?
4. Can you think of other ways that animals are able to escape their predators?

Science Buddies® Laura Candler • Kagan Publishing • 1-800-933-2667   **25**

Science Buddies Teacher's Page

Egg-citing Egg Trick

Overview

In this activity students soak an egg in vinegar and make an amazing discovery. The egg feels rubbery, but the inside remains runny. Egg shells are hard because they contain calcium. Vinegar is a mild acid which dissolves the calcium in the shell, leaving the thin membrane which surrounds the egg.

Introducing the Activity

Before handing out the Science Buddies activity, explain to students that they will be soaking an egg in vinegar. Have them think about what might happen and **RoundRobin** their predictions to their team.

Safety Notes: Remind students to wash hands thoroughly after handling the vinegar egg. Vinegar is an acid and must not come into contact with the eyes.

Cooperative Classroom Follow Up

On or after the activity due date, take a few minutes of class time to discuss the answers to Talk It Over. Use **Mix-Freeze-Pair** to involve the entire class. After discussing all the questions, tell the class that our teeth are made of calcium, just like the egg shell. For the last round of **Mix-Freeze-Pair** ask: "What might happen if we eat foods that contain acid and don't brush our teeth?" Explain that some nutritionists are concerned that chewable vitamins containing acetic acid (Vitamin C) may cause tooth decay.

Answers To Talk It Over

1. Students may have observed a foam on the surface of the vinegar, bubbles collecting on the egg, and a strong odor.
2. The egg will feel firm and hard before soaking in vinegar and somewhat like a balloon filled with water afterwards. When rubbed, the eggshell rubs off leaving a clear membrane. The egg looks yellowish at this point.
3. It's different because the vinegar dissolved the calcium out of the egg shell.
4. When held up to the light, the egg should look translucent; some light will pass through. The dark area is the yolk.
5. When pricked, the egg will burst and the inside will be runny. Many students are surprised because they think the egg will be firm and rubbery.

Science Buddies Activity

Egg-citing Egg Trick

Procedures

1. Place the raw egg in the glass or glass jar.
2. Cover the egg with vinegar.
3. Let the egg soak in vinegar for 24 hours. Observe the egg off and on during that time period.
4. After 24 hours, pour the vinegar out of the jar and carefully lift out the egg. Gently rub the eggshell. What happens?
5. Hold the egg up to a bright light. How does it look? Can you see through it?
6. What do you think will happen if you prick the egg with a needle or a pin? Place the egg back into the glass and try it!

Due Date

Materials

- 1 raw egg
- 1 cup vinegar
- Clear glass or jar
- Sharp object such as a pin or needle

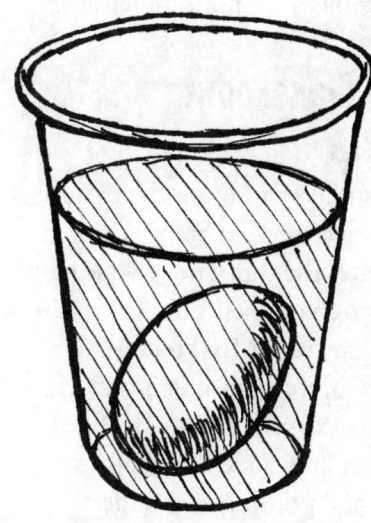

Talk It Over

1. What did you observe while the egg was in the vinegar?
2. How did the egg change after soaking in vinegar?
3. What do you think caused it to look and feel different?
4. How did the egg look when you held it up to the light?
5. Were you surprised by what happened when you pricked the egg with a pin?

Science Buddies® Laura Candler • Kagan Publishing • 1-800-933-2667

Science Buddies Teacher's Page

Tricky Twirler

Overview

In this activity students make an old-fashioned toy known as a thaumatrope. This toy was invented in 1825 and has fascinated people ever since. When the toy is twirled, a phenomenon known as "persistence of vision" causes the fish to appear to swim inside the fish bowl. Our eye retains an image for a fraction of a second after the actual image has disappeared. This ability is what allows us to view a motion picture or an animated cartoon as continually moving rather than as a series of still frames.

Introducing the Activity

Before handing out the Science Buddies activity, ask students to **RoundRobin** their favorite television cartoon. Then ask them to **Think-Pair-Share** their responses to these questions:
- *How are cartoons created?*
- *Why do the characters in cartoons seem to move when they are actually just a series of drawings?*

Tell students that their Science Buddies activity will help them understand how our eyes help us view cartoons and motion pictures.

Cooperative Classroom Follow Up

Remind students to bring the thaumatropes they created in Step 7 to class. On or after the due date, give students a chance to share their unique creations with their classmates. **Mix-Freeze-Pair** works well for this. Students mix around the room and then freeze on your signal. They pair with a partner and trade thaumatropes. After trying out and praising each others' Twirlers, students mix, freeze, and pair again. After students have shared their thaumatropes, have them return to their seats and respond to the questions using **Numbered Heads Together**.

Answers To Talk It Over

1. The fish appears to swim inside the fish bowl.
2. This optical illusion is caused by "persistence of vision," the eye's ability to retain an image for a fraction of a second after the image has disappeared. When the thaumatrope is twirled your eye sees both images at the same time causing the two pictures to blend together.
3. By observing thaumatropes we can easily see the phenomenon known as "persistence of vision." Cartoons also appear to move due to persistence of vision.

Science Buddies Activity

Tricky Twirler

Procedures

1. Carefully cut out the two circles below.
2. Trace the circles onto the index cards and cut out the two new circles.
3. Glue the circle with the fish onto one index card circle. Glue the fishbowl circle onto the other one.

Due Date

Materials

- Index cards (any size)
- Pencil (at least 6" long)
- Tape, glue, and scissors

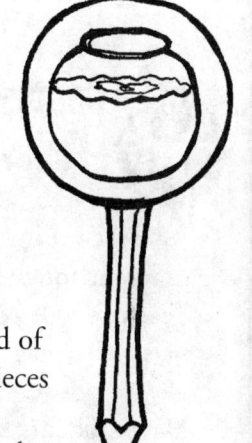

4. Place the circle with the fish FACE DOWN on the table. Put the end of the pencil on the back of the circle and tape it securely with several pieces of tape.
5. Place the other circle FACE UP over the end of the pencil. Tape the edges of both circles together carefully.
6. Hold the pencil between the palms of your hands and rub your hands back and forth quickly. Observe the picture in the circles.
7. You have just made an old-fashioned toy called a "thaumatrope." Can you create another thaumatrope which shows a different picture? (Bring your creation to class on the due date.)

1. What do you see when your twirl the pencil?
2. What do you think causes this illusion?
3. How can thaumatropes help us understand the way cartoons work?

Talk It Over

Science Buddies® Laura Candler • Kagan Publishing • 1-800-933-2667

29

Science Buddies Teacher's Page

Taste Test

Overview

This activity gives students a chance to explore their sense of taste. With our taste buds we can distinguish between four different tastes: bitter, salty, sour, and sweet. Taste buds in different locations on the tongue seem to be more sensitive to certain tastes than others. Generally, the back of the tongue can sense bitterness, the sides can detect sour and salty, and the front can taste sweet and salty.

Introducing the Activity

Before handing out the Science Buddies activity, give each student a small amount of two types of chocolate powder to taste. First have them taste unsweetened baking cocoa. Then have them taste chocolate drink mix powder. Have them **Think-Pair-Share** how the two powders are alike and different. Tell them that the science buddy activity will help them explore their sense of taste further.

Cooperative Classroom Follow Up

On or after the due date, use **Blackboard Share** to discuss the class results. Give each team a large sheet of construction paper. Within each team, assign an Artist to draw the tongue outline and a Recorder to write in the results of each teammate. Post each Team Tongue Map and compare the class findings to the results generally found by scientists. (See diagram.)

Have students **RoundRobin** their answers to the "Talk It Over" questions.

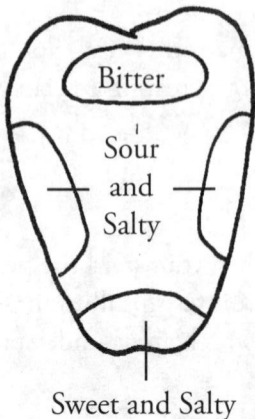

Answers To Talk It Over

1. Answers will vary.
2. Coffee tastes bitter, salt water tastes salty, sugar water tastes sweet, and vinegar tastes sour.
3. Answers will vary. Scientists have found that many people's tastes fall into the ranges shown in the illustration at right.

Science Buddies Activity

Taste Test

Procedures

1. Draw two tongue outlines similar to the ones below. Make each picture about 6 inches tall and 4 inches wide. Label the outlines with your names. You and your science buddy will taste several liquids and will use the drawings to "map" your taste buds.
2. Dip two cotton swabs in the strong black coffee (made without sugar). Give one to your science buddy so that you can test yourselves at the same time.
3. To perform the test, touch the swab to the front of your tongue. Then touch the swab to both sides and the back of your tongue. Where can you taste the coffee? Record the results on your tongue map.
4. Drink a little water to rinse out the coffee taste.
5. Test the vinegar, salt water, and sugar water in the same way. Be sure to rinse out your mouth before testing each liquid. Record your results on your individual tongue maps.

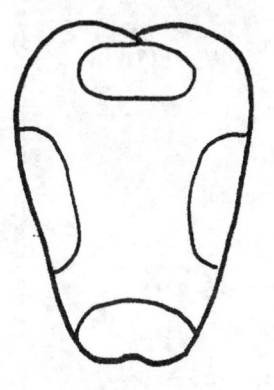

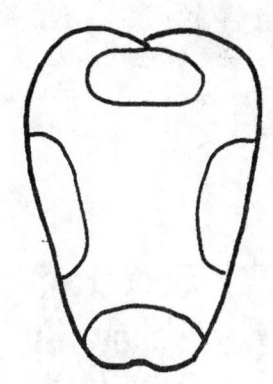

Due Date

Materials

- 8 cotton swabs
- 1 T. strong black coffee
- 1 T. vinegar
- 1 T. very salty water
- 1 T. very sugary water
- 2 glasses of water

My Tongue   Science Buddy's Tongue

Talk It Over

1. Did both of your tongue maps look the same?
2. Describe the taste of each liquid.
3. Could you taste all four liquids on all parts of your tongue? If not, where could you taste each liquid?

Science Buddies® Laura Candler • Kagan Publishing • 1-800-933-2667

Science Buddies Teacher's Page

Hot & Cold Race

Overview

This activity demonstrates how heat affects the motion of molecules. Molecules of hot water move around more rapidly than those of cold water. Dropping food coloring into the water shows the molecular action very clearly. Food coloring in hot water will spread throughout the water much more quickly than food coloring in cold water.

Introducing the Activity

Before handing out the Science Buddies activity, give each team a clear glass of lukewarm water. If possible, place a thermometer in each glass and have students read the temperature. Let them touch the water to feel its temperature. Drop a single drop of green food coloring into each glass and ask students to observe carefully. Then pose these questions for a **Team Discussion**:
- *Would food coloring behave differently in very cold or very hot water?*
- *What differences do you think you would observe?*

Cooperative Classroom Follow Up

On or after the due date, use **Mix-Freeze-Pair** to discuss the Talk It Over questions. Ask students to mix around the room and then freeze when you give a signal. Tell them to form pairs and discuss the first question. Call on several students to share their answers. Then have students mix, freeze, pair, and discuss the rest of the questions in the same manner.

Answers To Talk It Over

1. The drop of blue food coloring spreads SLOWLY throughout the cold water and the drop of red spreads QUICKLY throughout the hot.
2. Answers will vary.
3. The color should have spread evenly throughout both glasses after an hour.
4. The color of the food coloring is not important. Only the temperature of the water affects the action of the molecules.
5. The food coloring behaves differently because molecules of hot water move more quickly than those of cold water. Therefore, hot water molecules spread the food coloring throughout the glass more quickly than molecules of cold water.
6. Lemonade mix dissolves better in hot water due to the increased action of the water molecules. You might suggest dissolving the mix in a cup of hot water and then adding cold water and ice to fill up the container.

Science Buddies® Laura Candler • Kagan Publishing • 1-800-933-2667

Science Buddies Activity

Hot & Cold Race

Procedures

1. Pour 1/2 cup very COLD water into one glass. Have your Science Buddy pour 1/2 cup very HOT water into the other glass.
2. If you have a thermometer, measure the temperature of the water in each glass. Record your results.
3. Hold the bottle of blue food coloring over the glass of cold water. Ask your Science Buddy to hold the bottle of red food coloring over the hot water.
4. Predict what will happen when you drop food coloring into each glass. In which glass will the color spread faster?
5. On a count of three, both of you drop ONE drop of food coloring into your glass. *Do not stir!*
6. Watch both glasses closely for at least 5 minutes. Then observe them on and off for about an hour. Record your observations.

Due Date

Materials

- 2 identical clear cups or glasses
- 1/2 cup cold water
- 1/2 cup hot water
- Red and blue food coloring
- Thermometer (optional)

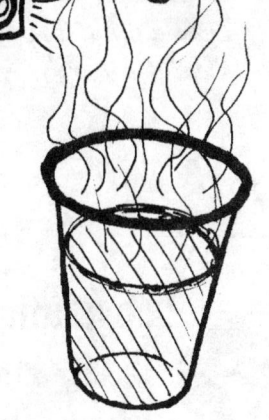

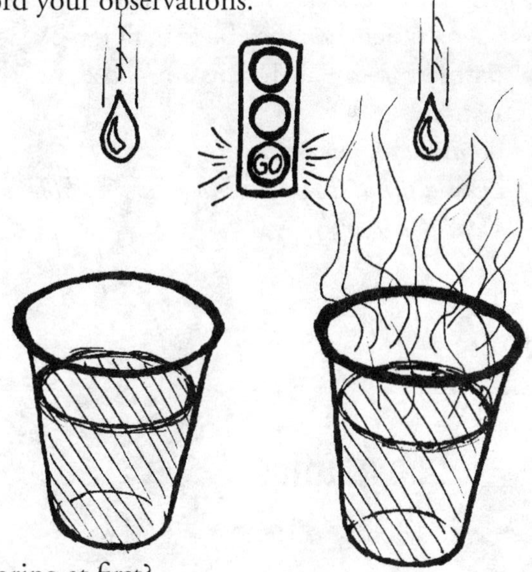

Talk It Over

1. What happened to each drop of food coloring at first?
2. Did you observe any other differences between the two glasses?
3. What happened to both drops after an hour?
4. What would happen if you used different colors of food coloring?
5. Why do you think the food coloring behaves differently in different water temperatures?
6. Do you think lemonade mix dissolves better in hot or cold water? Why? What would be a good way to make lemonade?

Science Buddies® Laura Candler • Kagan Publishing • 1-800-933-2667

Science Buddies Teacher's Page

Bubble Mania

Overview

In this activity, students will explore properties of water as they experiment with bubbles. Bubbles can not be blown with plain water because the surface tension, or force between water molecules, is too strong. Adding detergent to water weakens these cohesive forces and allows the soap film to stretch.

Advanced Preparation

No preparation is necessary since students can make their own bubble solution at home. However, you may want to make a batch for classroom use. Large batches of bubble solution can be made by adding 1 1/2 cups dish detergent to a gallon of cool water. Joy and Dawn seem to work best. You can also add a tablespoon of glycerin (obtainable at a drugstore) to make the bubbles last longer. Bubble-making works best on a humid day, since bubbles pop in dry air.

Introducing the Activity

Before handing out the Science Buddies activity, ask students to remember a time when they blew bubbles. Assign partners for a **Pair Discussion**. Pose these questions:
- *What shape are bubbles?*
- *Do you think it's possible to blow a bubble that's not round?*

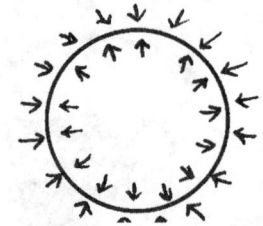

Cooperative Classroom Follow Up

Ask students to bring in their bubble makers on the due date. In **RoundRobin** fashion, have students show their bubble makers and share their experiences. Then use **Numbered Heads Together** to discuss the answers to the "Talk It Over" questions.

Answers To Talk It Over

1. All bubbles are round. The attraction between water molecules causes the soap film to pull together into the smallest possible shape. However, the air trapped inside the bubble is pushing out with equal force. A nearly perfect sphere is a result of these two forces acting together.
2. See explanation in Overview.
3. Answers will vary. Bubbles can be blown by making a circle with the thumb and forefinger and blowing through the opening. Very large bubbles can be made with a bent coat hanger or even a hula hoop! Plastic six-pack rings or strawberry containers can be used to make many bubbles at once.

Science Buddies Activity

Bubble Mania

Note: This activity should be completed outside since the bubble area will become wet and soapy.

Procedures

1. Pour the water into the bowl. Slowly add the dish detergent and stir gently. Try not to create any suds as you mix the water and detergent.
2. To make your first bubble-blowing devices, ask your Science Buddy to help you cut a hole in the bottom of each cup about the size of a dime. Both of you wet your hands and the cups with bubble solution since bubbles pop when they touch something dry.
3. Dip the large end of one cup into the bubble solution. Lift the cup out and gently blow into the hole in the small end. Now let your Science Buddy try. Who can blow the largest bubble?
4. To make your next bubble blower, twist the end of one pipe cleaner into a large loop. Use it to blow bubbles.
5. Try bending other pipe cleaners to form geometric shapes such as squares or triangles. Use them to blow bubbles also.

Due Date

Materials

- 2 tablespoons liquid dish detergent (Joy and Dawn work best)
- 1 cup cool water
- Flat-bottomed bowl
- 2 paper cups
- Scissors
- Pipe cleaners or wire
- Assorted items for blowing bubbles

Talk It Over

1. What shape are all your bubbles? Why do you think they form this shape?
2. Can you blow bubbles with plain water instead of the bubble solution? Try it. Can you explain the result?
3. What other bubble-blowing devices can you create? How could you make a very large bubble? How could you blow many bubbles at one time? Try out your ideas.

Science Buddies® Laura Candler • Kagan Publishing • 1-800-933-2667

Science Buddies Teacher's Page

Crystal Creations

Overview

Crystals fascinate children of all ages. In this activity, students learn about solutions and crystals by making homemade rock candy.

Introducing the Activity

Before handing out the Science Buddies activity, divide each team into two sets of partners for **Think-Pair-Share**. Pose the questions below and give students think time to determine their own answers. Then allow students to discuss their ideas with their partners. Finally, call on several students to share ideas with the class.

- *What would happen if you stirred sugar into water?*
- *How could you get the sugar back out of the solution?*
- *Would the sugar that came out of the solution still have the same properties as regular sugar?*

Safety Notes: Be sure to instruct students not to do this activity without adult supervision! The project involves heating a sugar and water mixture to the boiling point and pouring it into a glass, activities that create potential for harm to unsupervised children.

Cooperative Classroom Follow Up

On or after the due date, use **Team Discussion** to answer the "Talk It Over" questions. Follow each **Team Discussion** with a brief **Class Discussion** to clarify science concepts.

Answers To Talk It Over

1. Water can only dissolve a certain amount of sugar at room temperature. When the water molecules are heated they move quickly and allow a greater amount of sugar to be dissolved. The term for this type of solution is "supersaturated."
2. The sugar and water solution is different because it is thicker, stickier, and sweeter than plain water.
3. Crystals form because a supersaturated solution was created, which means that there is too much sugar for the water. When the solution cools to room temperature again, the crystals form as the sugar molecules collect on the popsicle stick. As the water continues to evaporate from the solution, more and more crystals form, creating rock candy.
4. The crystals taste sweet, showing that they have the same properties as sugar.

Science Buddies Activity

Crystal Creations

Note: Be sure to have an adult Science Buddy help you with this activity! Allow at least a week for crystals to form.

Procedures

1. Pour the sugar and the water into the pan and place it on the stove over medium heat. If you want colored crystals, add a few drops of food coloring now.
2. Stir gently as you bring the mixture to a boil. The sugar will not completely dissolve until the solution boils.
3. After the solution comes to a boil, remove the pan from the heat and let it cool at least 20 minutes.
4. Slowly pour the solution into the tall glass. Put the wooden skewer or popsicle stick into the glass.
5. Set the glass in a place where you can leave it undisturbed for at least a week. Spend a few minutes each day observing your crystals. If a crust forms over the solution, gently break the crust so the water can continue to evaporate.
6. When the crystals have grown as large as you like, remove the stick and examine them closely. You have made rock candy!

Due Date

Materials

- Measuring cup
- Pan
- Large spoon
- 1/2 cup water
- 1 1/4 cups sugar
- 2 clean popsicle sticks or wooden skewers
- Tall glass
- Tape
- Food coloring (optional)

Talk It Over

1. Why didn't the sugar dissolve until the solution boiled?
2. How is the solution different from the water you started with?
3. What caused the crystals to form?
4. How do the crystals taste?

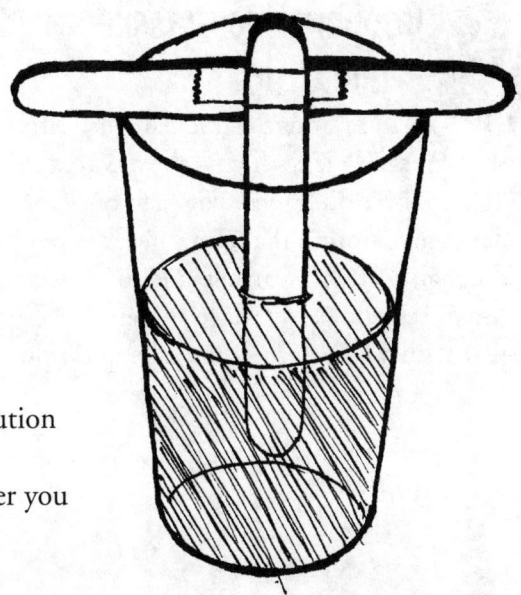

Science Buddies Teacher's Page

Ice Cream Investigations

Overview
Making homemade ice cream is a wonderful way to study matter. In this activity students make ice cream by placing a small ziplock bag with ice cream ingredients inside a large ziplock bag of ice and salt.

Introducing the Activity
Before handing out the Science Buddies activity, ask students to think of a special memory they have of eating ice cream. That memory might be a family celebration, a special flavor they enjoy, or sharing an ice cream sundae with a friend. Have students **RoundRobin** their memories to their teams. Then tell them that they will enjoy the chance to make their own ice cream at home with their Science Buddy.

Cooperative Classroom Follow Up
On or after the due date, use **Mix-Freeze-Pair** to discuss the answers to the Talk It Over questions. For each question students mix around the room, freeze when given a signal, pair with partner, and discuss their responses. Be sure to have students mix and pair with a new partner for each question.

Answers To Talk It Over

1. Students can change the flavor by substituting other extract flavors or adding flavored drink syrup (such as chocolate or vanilla).
2. The ice cream began as a liquid but became a solid at the end of the investigation.
3. The milk in the ice cream mixture froze and hardened.
4. Salt was needed because it lowers the freezing point of water. Pure water freezes at 32° Fahrenheit, but salt water freezes at 28° Fahrenheit. By lowering the freezing point, salt allows water to get colder than ice. This super-cold water and ice mixture causes the milk to freeze and become solid.
5. Salt is sprinkled on roads because it lowers the freezing point of water. The air temperature has to get below 28° to cause the water to freeze and form ice.

Science Buddies Laura Candler • Kagan Publishing • 1-800-933-2667

Science Buddies Activity

Ice Cream Investigations

Procedures

1. Pour the milk, vanilla extract, and sugar into the small ziplock bag. Squeeze out as much air as possible and seal the bag carefully.
2. Place the small ziplock bag down *inside* the large bag. Cover with the ice and salt. Seal the large bag tightly.
3. Take turns with your Science Buddy gently shaking, tossing, and flipping the two bags. If the bag gets too cold or starts leaking, wrap it with a towel.
4. Predict how long it will take for the liquid ice cream mixture to harden. Every minute or so, feel the small bag to see if the ice cream is ready. (Try to do this without opening the large bag.) Record the amount of time it took for your ice cream to harden.
5. When the ice cream is ready, open both bags and spoon the mixture into cups. Enjoy!

Due Date

Materials

- Measuring cup
- Measuring spoons
- Watch or clock
- 1 gallon ziplock bag
- 1 quart or pint ziplock bag
- 1 cup whole milk
- 1 teaspoon vanilla extract
- 3 tablespoons sugar
- 5 cups ice
- 1/4 cup salt
- Cups and spoons

Talk It Over

1. How could you change the flavor of your ice cream? Perhaps you can make another batch and try it!
2. What state of matter was the ice cream when you started the activity? What state of matter was it at the end?
3. What caused the ice cream to change states?
4. Why do you think the salt was needed?
5. Salt is often sprinkled on roads during the winter. Why do you think salt is used this way?

Science Buddies® Laura Candler • Kagan Publishing • 1-800-933-2667

Science Buddies Teacher's Page

Silly Slime

Overview

In this activity students create homemade slime, a fascinating substance that inspires endless exploration. Slime is a fluid, because a fluid is anything that flows. However, slime has some unusual characteristics that make it behave as both a solid and a liquid. It's actually a colloidal suspension, which means tiny bits of a solid (cornstarch) are floating, or suspended, in a liquid (water). Food coloring, an optional ingredient, adds to the fascination of this unusual substance.

Introducing the Activity

Before handing out the Science Buddies activity, place a paper towel in the center of each team. Spoon a small amount of cornstarch onto each paper towel and ask your students to examine it carefully. Before giving out the Science Buddies activity, pose the following **Team Discussion** questions:
- *What do you observe about the cornstarch?*
- *What do you think will happen if you mix the cornstarch with water?*

Safety Notes: The slime created in this activity is nontoxic, but discourage children from tasting it. The mixture will not be clean after extensive handling.

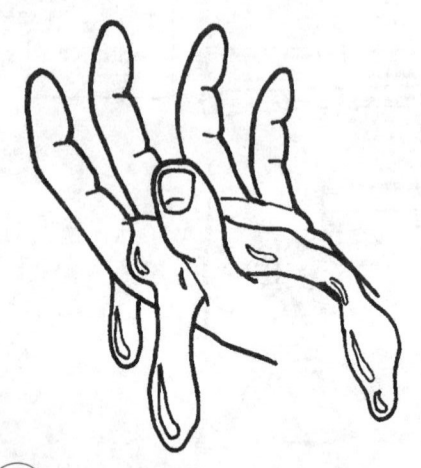

Cooperative Classroom

On or after the due date, discuss the activity as a class. Divide each team into pairs and use **Think-Pair-Share** to discuss the "Talk it Over" questions.

Answers To Talk It Over

1. It's both a solid and a liquid. (See explanation above.)
2. Answers will vary. Under pressure (as when squeezed or pounded) the slime will behave as a solid. When the pressure is released, the slime will flow like a liquid. When a tiny bit is rubbed between the fingers, it will feel powdery because the water evaporates leaving the cornstarch.

40 Science Buddies® Laura Candler • Kagan Publishing • 1-800-933-2667

Science Buddies Activity

Silly Slime

Note: This activity can be messy! Wear old clothes and perform it outside, if possible.

Procedures

1. Pour the water into the bowl. Add several drops of food coloring.
2. Slowly stir in about half the cornstarch. Keep adding cornstarch until the mixture is very thick and difficult to stir.
3. Quickly thrust the spoon into the slime. How does the slime feel? Now slowly push the spoon into the slime. How does it feel now?
4. Pour some slime into your hand. Can you roll it into a ball? Can you break the ball in half?
5. Spread your fingers apart and let the slime flow between them. Now quickly squeeze your hand into a fist. How does the slime feel?
6. Rub a small pinch of slime between your thumb and forefinger. How does the slime feel?
7. What else can you think of to do with your slime? Explore and experiment.
8. When you are finished with your slime, don't pour it down the sink. Scrape it into a plastic bag and discard it in the trashcan.

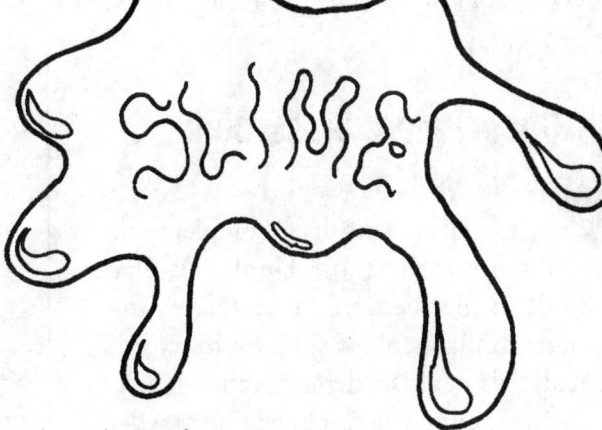

Due Date

Materials

- 1 cup cornstarch
- 1/2 cup water
- Food coloring (optional)
- Small bowl
- Metal or wooden spoon
- Plastic bag

Talk It Over

1. Do you think slime is a solid or a liquid? Explain your answer.
2. What was the most interesting discovery you made with your slime?

Science Buddies® Laura Candler • Kagan Publishing • 1-800-933-2667

Science Buddies Teacher's Page

Cabbage Juice Chemistry

Overview

Cabbage juice is one of many natural indicators. An indicator is a material that changes color in the presence of another substance, in this case an acid or a base. Cabbage juice turns **green** in a base and **red** in an acid. In this activity students become "chemists" as they explore the effects of household liquids on cabbage juice.

Introducing the Activity

Before handing out the Science Buddies activity, ask students if they have ever been in a swimming pool. Have them **RoundRobin** ways that the water in a pool is different from ocean or pond water. Next use **Team Discussion** to talk over the steps that a lifeguard might take to keep the water clean and safe for swimming. Students may not be aware that chemicals such as chlorine and hydrochloric acid are added to water. Lifeguards and pool owners use test kits to determine the pH of the water. The test kits contain liquids that change color if the water is too acid or basic. These liquids are called indicators. Explain to your students that the Science Buddies activity will allow them to discover more about indicators.

Safety Notes: Caution students against trying other household liquids in this investigation. Many cleaning products give off harmful fumes. Furthermore, some liquids (such as bleach and ammonia) can give off deadly fumes when mixed.

Cooperative Classroom Follow Up

After all students have completed the activity at home, use **Think-Pair-Share** to discuss the questions. If you have cabbage juice available, allow students to try other safe materials in a class demonstration (such as a crushed vitamin tablet, salt, baking powder, milk, or other juices).

Answers To Talk It Over

1. Orange juice, vinegar, and aspirin are acids because they cause cabbage juice to turn red in color. Baking soda and shampoo are bases because they turn cabbage juice green. (Some "pH balanced" shampoos may be neutral.)
2. Students should have observed that even though they mixed equal amounts of acids and bases into the cabbage juice, the color changes were not consistent. The stronger the acid or base, the deeper the red or green color.

42 Science Buddies® Laura Candler • Kagan Publishing • 1-800-933-2667

Science Buddies Activity

Cabbage Juice Chemistry

Procedures

Note: You will need to prepare "cabbage juice" for this activity. This special liquid called an <u>indicator</u> changes color in the presence of an acid or a base. Caution: Do not prepare the liquid without adult supervision!

Materials

- 1/2 head red cabbage
- 3 cups water
- Large pot
- 5 clear glasses or cups
- 1 tsp. each: lemon or orange juice, baking soda, vinegar, shampoo
- 1 aspirin tablet (optional)

1. Have your Science Buddy help you make the cabbage juice several hours (or even a day) ahead of time. Follow these steps to make the cabbage juice:
    - Carefully chop the red cabbage into small pieces.
    - Place the pieces into a pot and cover with 3 cups water.
    - Bring to a boil, then turn off the heat and allow the mixture to stand several hours until cool.
    - Strain off and discard the cabbage. Refrigerate the purple liquid.
2. Pour 1/4 cup cabbage juice into each of the 5 cups.
3. Add the lemon juice to the first cup. Record the color change.
4. Add the baking soda to the second cup, the vinegar to the third cup, and the shampoo to the fourth cup. Record each color change.
5. Crush the aspirin tablet with a spoon and add it to the last cup. Record the color change.

Talk It Over

1. Lemon juice is an acid. Which of the materials you tested were acids and which were bases. Why?
2. What evidence did you observe that some acids and bases are stronger than others?

Science Buddies® Laura Candler • Kagan Publishing • 1-800-933-2667

Science Buddies Teacher's Page

Raisin Razzmatazz

Overview

Raisins appear to come to life in this activity, zipping up and down in a mixture of vinegar and baking soda. The investigation can be used to extend a unit on physical and chemical changes or to supplement a lesson on density and properties of matter. The activity also serves to reinforce the process skill of observation.

Introducing the Activity

Before handing out the Science Buddies activity, use **Pair Discussion** to explore the concepts of floating and sinking. Before distributing the Science Buddies activity page, have students talk over the following questions with their partner:
- *What kinds of things float?*
- *Is it possible for large, heavy objects to float?*
- *Can small objects sink?*
- *Do you think a raisin would sink or float?*

Cooperative Classroom Follow Up

On or after the due date, discuss the Talk It Over questions. First ask each team member to **RoundRobin** their predictions for the activity. Then use **Numbered Heads Together** to discuss the answers to questions 2 - 5. If many students tried other objects and liquids, use **Blackboard Share** to post their results. Each team prepares a chart similar to the one shown to share with the class.

Objects	Liquids

Answers To Talk It Over

1. Answers will vary.
2. The raisins should bob up and down, floating to the top and then sinking to the bottom of the glass.
3. The chemical reaction between vinegar and baking soda causes bubbles of carbon dioxide gas to be released. The bubbles cling to the wrinkles in the raisins, causing the raisins to become less dense than the water and float. When the raisins reach the top the bubbles pop allowing the raisins to sink.
4. Answers will vary. Objects must be slightly heavier than water and have a rough texture that allows bubbles to collect on the surface.
5. Answers will vary. Carbonated drinks also work for this activity.

44   Science Buddies® Laura Candler • Kagan Publishing • 1-800-933-2667

Science Buddies Activity

Raisin Razzmatazz

Procedures

1. Place the glass on the plate or tray. Measure the water and pour it into the glass. Gently stir in the baking soda.
2. SLOWLY pour the vinegar into the glass of baking soda and water. Observe carefully. (If the solution overflows, empty the glass and start over.)
3. Predict what will happen when you drop the raisins into the glass.
4. Drop the raisins into the cup and observe carefully. (If nothing happens, remove the raisins and try new ones. This time cut the raisins in half before dropping them into the liquid.)
5. Observe the raisins for at least 10 minutes. Write a list of your observations.

Due Date

Materials

- Clear glass or cup
- Spoon
- Tray or plate
- 1 tablespoon baking soda
- 1/4 cup water
- 1/2 cup vinegar
- 5 raisins

Talk It Over

1. What did you predict would happen to the raisins?
2. How did the raisins behave in the vinegar and baking soda solution?
3. Why do you think the raisins behaved as they did?
4. What else could you use in this experiment instead of the raisins?
5. What other liquids might work instead of the baking soda and vinegar? Try out some of your ideas and record your results.

Science Buddies® Laura Candler • Kagan Publishing • 1-800-933-2667

Science Buddies Teacher's Page

Creeping Colors

Overview

"Creeping Colors" involves paper chromatography. The word "chromatography" comes from the Greek term for "color writing." It's used in analytical chemistry to separate and identify the components of mixtures. Using paper filters is the simplest form of chromatography, though this method is not commonly used in science today. Modern science uses a variety of other chromatography methods. After students experiment with coffee filters and colored makers, you may want them to research other chromatography techniques.

Introducing the Activity

Before handing out the Science Buddies activity, assign the following question for a **Team Discussion**: *"What are some ways that you could separate and identify the parts of an unknown mixture?"* Use **Blackboard Share** to allow teams to present their ideas to the class. Give each team a piece of construction paper and a marker. Assign a Recorder to write down the team's ideas and a Reporter to present them to the class.

Cooperative Classroom Follow Up

On or after the due date, allow students to show each other their coffee filters. **RoundRobin** works well for this. Team members take turns standing for 30 seconds and showing their filters. They explain how they obtained their results, including colors and brands of markers used. Use **Numbered Heads Together** to discuss the Talk It Over questions.

Answers To Talk It Over

1. The water climbs up the paper because of *capillary action*, the attraction of water molecules to the tiny openings in the fibers of the coffee filter.

2. Water-soluble inks often contain several different colors which are blended to form the visible ink color. As the water moves through the ink mark, it dissolves the ink and separates the various colors. The lighter components are carried farther up the filter paper.

3. Dictionary definitions may vary. Chemists use chromatography to separate mixtures. Biologists use chromatography to analyze blood and tissue. Ecologists use such procedures to identify small amounts of pollutants in air, water and food.

Science Buddies® Laura Candler • Kagan Publishing • 1-800-933-2667

Science Buddies Activity

Creeping Colors

Procedures

1. Smooth out one of the coffee filters. Cut a 1 inch wide strip from the edge of the filter to its center as shown.
2. With one of the markers, make a horizontal line across the edge of the strip about 1 inch from the bottom.
3. Half fill the glass with water.
4. Place the coffee filter on top of the glass so the strip hangs down into the water as shown. The mark should just **above** the water. You may have to adjust the water level.
5. Observe the water as it "climbs" up the paper.
6. After 10 minutes, remove the filter and let it dry.
7. Repeat steps 1-6 using clean coffee filters and at least two other markers. Try different brands of markers if you can.
8. Bring your dried coffee filters to class on the due date.

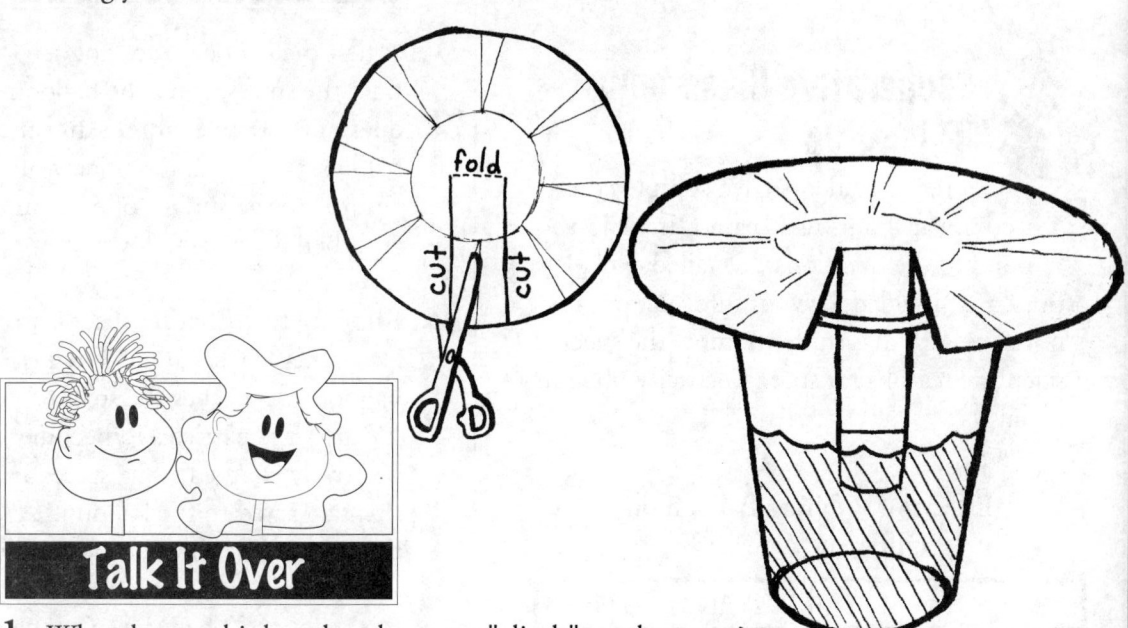

Due Date

Materials

- 3 or more clean white coffee filters
- Washable markers
- Clear drinking glass
- Water

Talk It Over

1. What do you think makes the water "climb" up the paper?
2. What happened to the marker colors?
3. The scientific name for what you have done is "paper chromatography." Look up "chromatography" in the dictionary. What does it mean? When might scientists use chromatography?

Science Buddies® Laura Candler • Kagan Publishing • 1-800-933-2667

Science Buddies Teacher's Page

Mysterious Balloon

Overview

Balloons provide a dramatic way of demonstrating static electricity. When a balloon is rubbed against certain materials, electrons are given up by that material and collect on the balloon. This transfer of electrons causes the balloon to be negatively charged. Since items with "like" charges repel each other, and items with "unlike" charges attract, the balloon will attract objects with a positive charge.

Introducing the Activity

Before handing out the Science Buddies activity, blow up a balloon and rub it vigorously against your hair or clothing. Hold the balloon up to the wall and let go. The balloon should cling to the wall. (Static electricity demonstrations work best in a dry environment.)

Assign partners and use **Pair Discussion** to talk over reasons why the balloon sticks to the wall. Do not give the "correct answer;" just tell them that they will investigate this mysterious property during the Science Buddies activity.

Cooperative Classroom Follow Up

After all students have completed the activity at home, use **Team Discussion** to respond to the questions. Detailed explanations are provided for your information; it's not necessary that students know the specific scientific reasons for static electricity phenomenon.

Effects of Rubbing Balloon on Different Materials

	Shirt	Pants	Hair
Pepper			
Basil			
Oregano			
Thyme			

Answers To Talk It Over

1. At first the balloon does not attract the spices, since the balloon does not carry a charge. After the balloon is rubbed, the spices will "jump" to the surface of balloon and then drop back down after a few seconds.
2. Rubbing the balloon causes it to have a negative charge. When the balloon approaches the spices, it causes the particles to become positively charged. The spices are then attracted to the balloon. As the spices cling to the balloon, they slowly drain electrons away from the balloon. When the spices have a neutral charge, they fall back to the paper.
3. Answers will vary. See sample chart.

Science Buddies Teacher's Page

48 Science Buddies® Laura Candler • Kagan Publishing • 1-800-933-2667

Science Buddies Activity

Mysterious Balloon

Note: This activity works best on a dry day.

Procedures

1. Blow up the balloon and tie it off.
2. Draw four large circles on the piece of paper.
3. Sprinkle 1/4 tsp. of one spice in each circle. Write the name of the spice under each circle.
4. Hold the balloon about an inch above each circle, one at a time. Observe carefully.
5. Rub the balloon firmly back and forth against your shirt or pants for about 15 seconds. What do you think will happen when you hold the balloon above the spices this time?
6. Test your predictions. Move the balloon over the circles. Observe carefully for several minutes.
7. What else can you discover about your mysterious balloon? Will this activity work with different spices? How about rubbing the balloon against different fabrics? Try out some of your ideas.

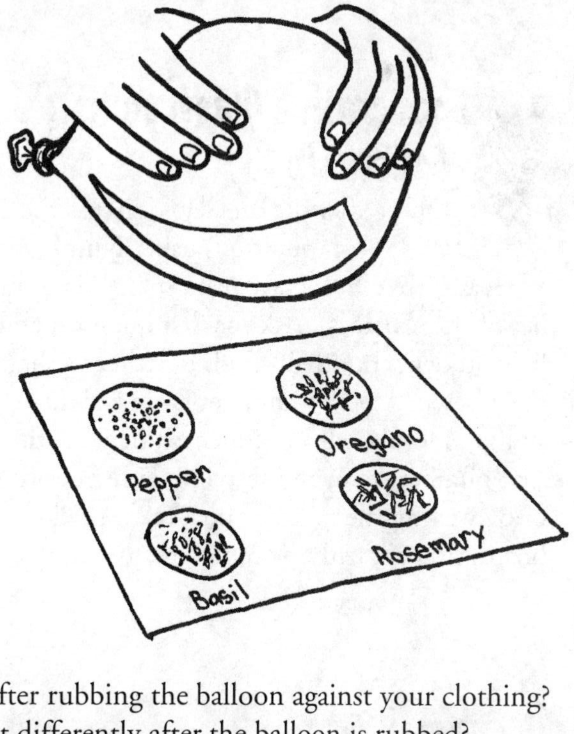

Due Date

Materials

- 1 balloon
- 1/4 tsp. of four spices (pepper, oregano, basil, thyme, etc.)
- Plain white paper

Talk It Over

1. How did the spices behave before and after rubbing the balloon against your clothing?
2. What do you think makes the spices act differently after the balloon is rubbed?
3. How can you make a chart of your investigation results?

Science Buddies® Laura Candler • Kagan Publishing • 1-800-933-2667

49

Science Buddies Teacher's Page

Sun Fun

Overview
This activity introduces students to the power of solar energy. By placing zip lock bags of water in direct sunlight on black and white paper, students discover that dark colors absorb heat better than light colors.

Introducing the Activity
Before handing out the Science Buddies activity, ask students to **RoundRobin** their responses to this question: *"What is your favorite way to keep cool on a hot summer day?"* Afterwards, tell them that this month's Science Buddies activity will give them additional ideas for keeping cool in the sun.

Cooperative Classroom Follow Up
On or after the due date, discuss the Talk It Over questions using **Numbered Heads Together**. Have team members number off from 1 to 4. Ask the first question and have students put their heads together to share their ideas. After a minute or two of discussion, randomly call a number. Students with that number stand and respond, either orally or by writing their answer on a team chalkboard. Repeat with the remaining questions.

Answers To Talk It Over

1. Heat energy from the sun warms the water in both bags.
2. The water on the black paper becomes noticeably warmer because dark colors absorb more heat energy than light colors. The black paper warms quickly and the heat energy is transferred to the water in the bag.
3. Other colors will affect the temperature, but the effect is not as noticeable. Dark paper will result in warmer temperatures than light colors.
4. White T-shirts are the coolest on a sunny day because they reflect the sunlight instead of absorbing it.
5. Answers will vary, but may include the following:
   • Choosing paint color for a house in a warm or cold climate.
   • Selecting a color for the interior or exterior of a car.
   • Deciding whether to step on a blacktop road or a white concrete sidewalk.
   • Planning clothes for a vacation.

Science Buddies® Laura Candler • Kagan Publishing • 1-800-933-2667

Science Buddies Activity

Sun Fun

Note: This experiment must be done on a bright, sunny day.

Procedures

1. Pour 1/2 cup cold water into each zip lock bag. If you have a thermometer, measure the temperature of the water in each bag and record it on a piece of paper.
2. Seal both bags tightly. Make sure neither bag leaks.
3. Place the two sheets of paper on a flat surface in direct sunlight. Lay one zip lock bag of water on the black paper and the other bag on the white paper.
4. Leave the two bags in the sun for 30 minutes. What do you think will happen to the water temperature in each bag?
5. After 30 minutes, pick up each bag and feel their water temperatures. Which bag is warmer?
6. If you have a thermometer, measure and record the temperature of the water in each bag. What is the difference in water temperature between the two bags?

Due Date

Materials

- 1 cup water
- 2 small zip lock bags
- 1 sheet black paper
- 1 sheet white paper
- Measuring cup
- Thermometer (optional)
- Clock or watch

Talk It Over

1. What caused the water temperature in
2. Why was one bag warmer than the other one?
3. Do you think other colors of paper would affect the water temperature? Try another experiment to find out!
4. What color T-shirt would be the coolest to wear on a hot, sunny day?
5. Knowing about the effects of dark and light colors might be important in other situations. Can you think of any examples?

Science Buddies® Laura Candler • Kagan Publishing • 1-800-933-2667

Science Buddies Teacher's Page

Color Spinners

Overview

In this activity students create a simple toy which allows them to explore the concepts of light, color, and optical illusions. The black and white spinner demonstrates an amazing phenomenon: repeated flashes of light cause the retina to see color where none exists. Other interesting color wheels can be made with primary colors which appear to blend and create secondary colors.

Advanced Preparation

Duplicate one set of Color Spinner Patterns for each student. Make a sample spinner from one of the blank patterns to demonstrate the proper twirling technique. Do not color the spinner since that may spoil the fun of discovery.

Introducing the Activity

Before handing out the Science Buddies activity, put a glass of water in the center of each team. Drop yellow and blue food coloring into the glass and ask students to observe carefully. They will quickly notice that the water begins to look green.

Pose this question for **Team Discussion**: *"What are some other ways you can mix colors?"* Tell them that they will explore color mixing in their Science Buddies activity. You may also want to show them your blank Color Spinner and demonstrate how to twirl it.

Cooperative Classroom Follow Up

Set a date for students to bring their color spinners to class. Allow students to share their spinners with **Mix-Freeze-Pair**. Have students mix around the room, and then freeze when you give a signal. Ask them to pair with a partner and trade spinners. Remind students to praise each other for their creativity, color choices, and artistic design. Then have them mix, freeze, and pair several more times so they can test many different spinners. Have students return to their seats and use **Think-Pair-Share** to discuss the answers to the Talk It Over questions.

Answers To Talk It Over

1. The black and white pattern appears to have color, due to an optical illusion. Repeated flashes of white light cause our eyes to see color when none exists.
2. The blue and yellow appear green. This happens because the colors are moving so quickly that the eye cannot keep them separate and blends them together in an optical illusion.
3. A spinner that appears orange can be created by coloring the spinner red and yellow.

Science Buddies Activity

Color Spinners

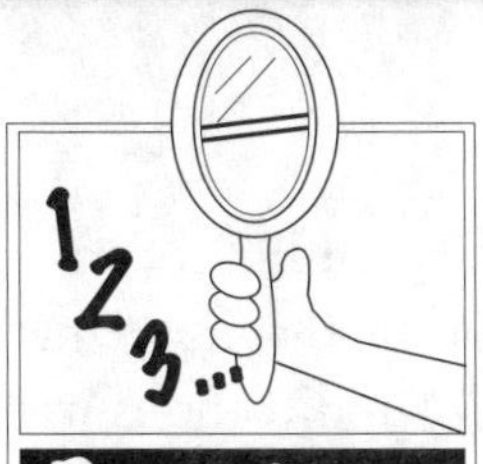

Procedures

1. Cut out the black and white spinner. Trace the spinner onto the sheet of posterboard. Cut out the circle of posterboard and glue the spinner onto it.
2. Have your Science Buddy use the nail or paper clip to punch two small holes near the center of the disk where shown.
3. Thread the string through the holes and tie the ends together to make a long loop.
4. Predict what you will see when you spin the disk.
5. Hold the ends of the loop with the spinner centered in the middle. Wind up the disk and make it twirl by tightening and relaxing the string. (This takes a bit of practice, so keep trying.) Observe the spinner carefully. What do you see?
6. Try making a color spinner with one of the blank patterns. Color this one blue and yellow (alternate colors). Predict what will happen when you twirl it. Test your prediction by spinning it.
7. Make up your own color spinner with a third pattern, and let your Science Buddy create one with the last pattern. Who can make the most interesting spinner?

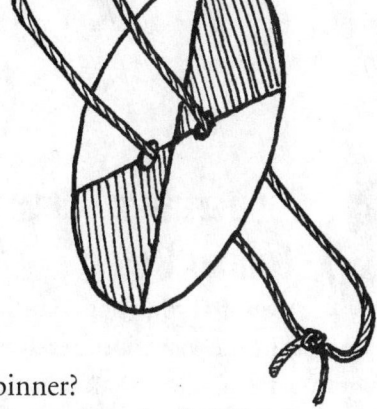

Due Date

Materials

- Color Spinner Patterns
- Posterboard or cardboard
- 1 yard of string or yarn
- Scissors
- Pencil
- Crayons or markers
- Glue
- Nail or paper clip

Talk It Over

1. What did you see when you twirled the first color spinner?
2. What did you see when you twirled the blue and yellow color wheel? Why do you think this happened?
3. How could you make a spinner that looks orange when twirled, without using the color orange anywhere on the spinner?

Science Buddies® Laura Candler • Kagan Publishing • 1-800-933-2667

Science Buddies Teacher's Page

The Magic Card

Overview

This simple trick involves filling a cup with water and placing an index card on top. When the cup is inverted, the card clings to the cup as if by magic. Actually, the force of air pressure is at work. The air around us is pushing in all directions with a force of about 15 pounds per square inch. (We don't feel this air pressure because of the equal air pressure inside our bodies.) The force of the air is actually greater than the weight of the water, so the card remains in place.

Introducing the Activity

Before handing out the Science Buddies activity, ask students to think of a time when they had trouble pouring something out of a can (such as soda or milk). If possible, demonstrate by opening a can of soda and turning it directly upside down to pour it in a glass. The soda will "chug" out instead of pouring smoothly. Have students turn to a partner for a **Pair Discussion** of what causes this problem and what ideas they have for solving it. Don't try out their solutions; just encourage them to keep the problem in mind as they are doing their Science Buddies activity.

Safety Notes: Remind students to be careful not to cut themselves on the sharp edge of the paper clip when punching the hole.

Cooperative Classroom Follow Up

After everyone has completed the activity at home, use **Numbered Heads Together** to discuss the answers to the questions.

Answers To Talk It Over

1. In the first experiment the card hangs in the air as if by magic. After the hole is poked in the bottom of the cup, the water and the card quickly fall into the sink.

2. In the first experiment, air pressure keeps the card up. During the second experiment, air goes into the cup through the hole and pushes with equal force down on the water and card.

3. By observing the how air pressure affects water in a cup, students should be able to pour something out of a can more easily. Hopefully they learned that for a liquid to flow out of a can, air must be able to flow in at the same time. This can be achieved in several ways. A second hole can be punched into a can, if convenient. Also, the can may be held in a more horizontal position so that air can flow in at the same time the liquid is flowing out.

Science Buddies® Laura Candler • Kagan Publishing • 1-800-933-2667

Science Buddies Activity

The Magic Card

Procedures

1. Do this activity over a bucket or sink.
2. Fill the cup with water.
3. Place one index card on top of the cup so that the opening is completely covered.
4. With one hand on the cup and the other on the card, quickly turn the cup upside down.
5. Take your hand away from the card. Observe carefully.
6. Pour out the water. Open the paper clip slightly and use the end to poke a small hole in the bottom of the cup.
7. Cover the hole with your finger while you fill the cup with water again. Use a dry index card to cover the mouth of the cup.
8. With your finger still over the hole, hold the index card in place and turn the cup upside down.
9. Predict what will happen when you take your finger away from the hole. Test your prediction.

Due Date

Materials

- Plastic, styrofoam, or paper cup
- 2 index cards or pieces of thin cardboard
- Bucket or sink
- Water
- Paper clip

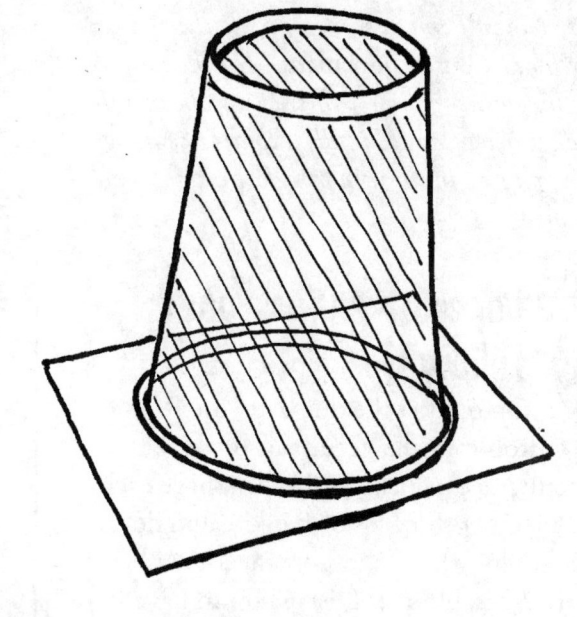

Talk It Over

1. What happens to the card in both experiments?
2. Why do you think the card behaves differently in each situation?
3. What did you learn that might help you pour liquid out of a can more easily?

Science Buddies® Laura Candler • Kagan Publishing • 1-800-933-2667

Science Buddies Teacher's Page

Amazing Paper Kite

Overview

In this activity students make a simple paper kite out of a sheet of paper, strips of tissue, and a spool of thread. Despite the simplicity of the kite, it will fly almost as well as a manufactured kite. While flying their kites, students can observe basic forces involved in flight: *lift* caused by the wind, the pull of *gravity*, and *drag* (air resistance).

Advanced Preparation

Before handing out the Science Buddies activity, duplicate one "Amazing Paper Kite" pattern for each student, making sure the dashed line is centered exactly. If possible, follow the directions to assemble one sample kite in advance for students to examine during the lesson introduction.

Introducing the Activity

Before handing out the Science Buddies activity, ask students if they have ever flown a kite or observed someone flying a kite. Have them **Think-Pair-Share** their responses to the question: *"What makes a kite fly?"* Tell them that they will be making and flying their own kites. If you have prepared a sample kite to show them, let students examine it at this time. Give them each one pattern and ask them to bring their paper kites to class on the due date.

Safety Notes: Remind students to fly their kites away from streets or busy roads. Tell them to wear sunglasses to protect their eyes if they fly the kite on a bright day. Finally, explain the dangers of flying a kite in the rain or before an approaching storm.

Cooperative Classroom Follow Up

On or after the due date, use **RoundRobin** to allow students to share their kite-flying experiences. In turn, have each team member show his or her kite and describe when, where, and how the kite was flown. Next, use **Pair Discussion** to have students respond to the questions. Finally, if weather and space permit, take the class outside for a kite-flying celebration!

Answers To Talk It Over

1. The straw is needed to stabilize the wings. It keeps the wings from flapping loosely and allows the wind to be trapped underneath. The tail pieces are needed to create more drag and hold the kite steady in strong winds.
2. A kite flies because of the force of lift created by the wind. Lift must be strong enough to overcome the force of gravity which is pulling the kite downward.

Science Buddies Activity

Amazing Paper Kite

Note: See the illustrations below for each step.

Procedures

1. Fold the Kite Pattern in half on the dashed line. Then fold each wing of the kite down on the dotted lines.
2. Lift the wings to the flat position as shown. Tape across the folded edges in two places to hold the kite together. Tape a straw on the top side of the kite about 1 inch from the front edge (above black dots).
3. With the paper clip, poke a hole through the two dots on the bottom edge of the kite. *Leaving the thread attached to the spool*, thread the free end through the hole. Tie the thread to itself. Decorate your kite with crayons, markers, or stickers.
4. To make the tail, cut 4 strips of tissue paper about 1 1/2 inches wide and 4 feet long. Tape each piece to the upper back edge of the kite.
5. Fly your kite on a breezy day. Hold the kite up and run into the wind. As the kite catches a breeze, let out the thread. You may have to experiment with the lengths of the tail pieces depending on the amount of wind present.

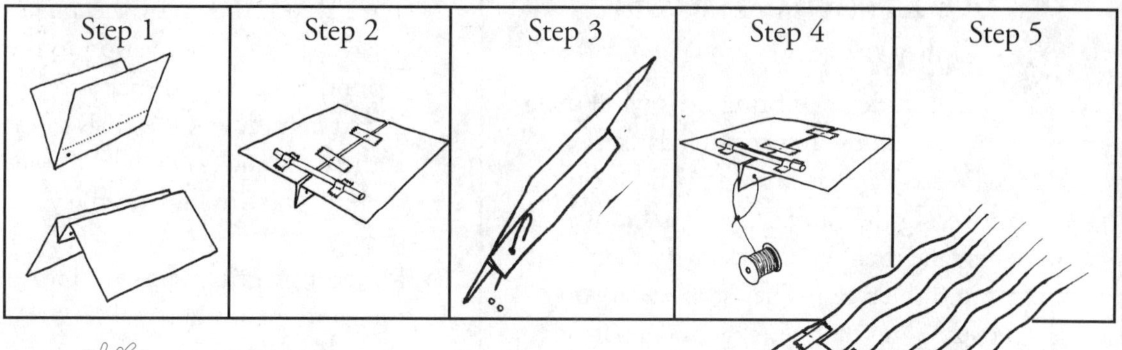

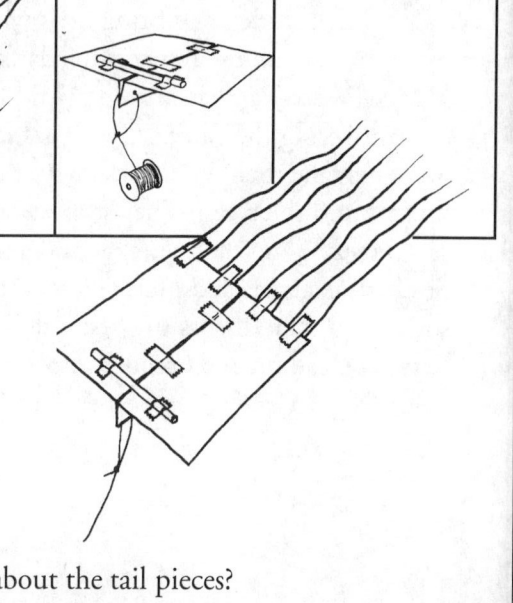

Due Date

Materials

- 1 Kite Pattern
- Scissors
- 1 spool of thread
- Tissue paper, crepe paper, or bathroom tissue
- 1 straw
- 1 paper clip

Talk It Over

1. Why do you think the straw is needed? What about the tail pieces?
2. What did you discover about flying a kite? What makes it fly?

Science Buddies® Laura Candler • Kagan Publishing • 1-800-933-2667

Science Buddies Teacher's Page

Wonderful Wind Socks

Overview
The study of wind is critical to the science of meteorology. In this activity, students create wind socks and make observations for several days.

Introducing the Activity
Before handing out the Science Buddies activity, ask the following questions and have students **RoundRobin** their answers to their teammates:
- *How can you tell the direction the wind is blowing?*
- *How can you tell how fast the wind is blowing?*
- *Why do scientists study the wind?*

Cooperative Classroom Follow Up
Ask students to bring their wind socks to class on or after the due date. Use **Mix-Freeze-Pair** for showing wind socks and answering the Talk It Over questions. Students take their wind socks with them and mix around the room. They stop when you say "Freeze!" and then quickly find a partner. Students examine and praise each others' wind socks, and then discuss the first question. After you call on a few students to respond aloud, say "Mix!" and repeat the activity.

Answers To Talk It Over

1. Students will observe the streamers blowing to one side and, in a stronger breeze, the top of the wind sock moving also.
2. The wind direction is shown by the direction the streamers move. (Looking at a compass will show you the true direction.)
3. The speed is shown by the angle the wind sock is in relation to the ground. If the wind sock is hanging straight down, the air is calm. If the wind sock is parallel to the ground, the wind is extremely strong.
4. Other instruments used to measure the wind include a *wind vane* to show direction and an *anemometer* to indicate wind speed.
5. Answers will vary. Winds indicate approaching weather and potential weather changes. Studying wind is also important for aircraft pilots. Architects study the affects of wind on tall buildings so that they can create structures that are flexible yet strong.

Science Buddies Teacher's Page

Science Buddies® Laura Candler • Kagan Publishing • 1-800-933-2667

Science Buddies Activity

Wonderful Wind Socks

Procedures

1. To make the top of the wind sock, cut a large piece of heavy paper into a 6" x 18" strip.
2. Decorate the paper strip with crayons or markers.
3. To make streamers, cut six strips of crepe paper or tissue 2" wide and 30" long.
4. Lay the top of the wind sock face down and tape the streamers to the bottom edge. Space the streamers about 1" apart.
5. Roll the paper strip into a tube and tape the edges of the wind sock together.
6. Punch two holes directly across from each other in the top edge of the wind sock.
7. Cut a 24" long piece of string or yarn. Tie the yarn through the two holes to form a handle.
8. Hang your wind sock on a tree branch, or just hold it out in the breeze. What do you observe?
9. Take your wind sock out several times over the next few days and continue to make observations.

Due Date

Materials

- Construction paper or paper grocery bag
- Crepe paper streamers, tissue paper, or bathroom tissue
- Tape
- Yarn or string
- Crayons or markers
- Scissors
- Hole puncher or nail

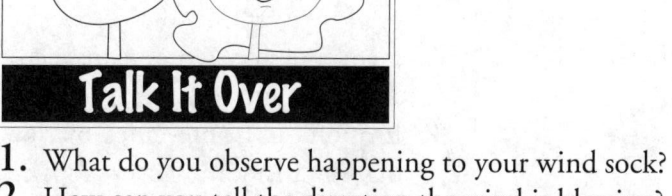

Talk It Over

1. What do you observe happening to your wind sock?
2. How can you tell the direction the wind is blowing?
3. How can you tell how fast the wind is blowing?
4. What other weather instruments measure wind speed and/or direction?
5. Why is it important to study the wind?

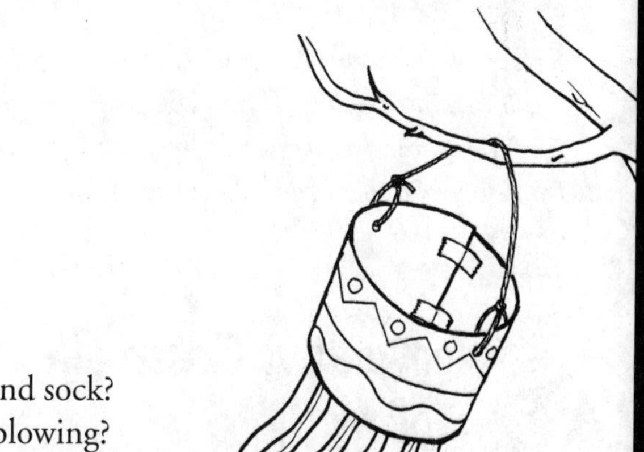

Science Buddies Teacher's Page

Weathering & Erosion Walk

Overview
Weathering and erosion takes place in every outdoor environment, from the most urban playground to the most rustic farmyard. In this activity students take a walk with their Science Buddies, observing and recording signs of weathering and erosion in their own neighborhoods. (Weathering and Erosion Log found on Page 78).

Introducing the Activity
Before handing out the Science Buddies activity, use **Pair Discussion** to talk over the following:
- *How do you think sand and soil are formed?*
- *How are sand and soil moved from one place to another in nature?*

Make sure students understand that *weathering* is the breaking down of rocks into smaller pieces, whereas *erosion* is the movement of those pieces by wind or water. Soil contains bits of rock as well as organic material such as decayed plant matter.

Safety Notes: *Remind students to walk with an adult Science Buddy. If they visit construction sites they should wear sturdy shoes and refrain from picking up discarded materials.*

Cooperative Classroom Follow Up
Use **RoundRobin** for sharing the Weathering and Erosion Logs. Students take turns within their teams holding up their Logs and explaining their discoveries. Then use **Think-Pair-Share** to discuss the activity questions.

Answers To Talk It Over
1. See explanation at left.
2. Erosion frequently occurs where water is able to run off, such as stream beds, steep hillsides, and construction sites. Wind erosion may be an important factor in some areas, such as beaches, open farmland, and deserts.
3. Problems include: sidewalks cracked by roots and grass, potholes created by freezing and thawing water, loss of soil by farmlands, muddy rivers due to excessive sediment, marble statues damaged by acid rain, and beach erosion.
4. Erosion can be prevented by slowing water runoff. Cover crops can be grown during the winter, grass can be planted on construction sites, and farmers can plow across a hill rather than up and down the slope. Wind erosion can be prevented by planting rows of trees to slow the wind.

Science Buddies Laura Candler • Kagan Publishing • 1-800-933-2667

Science Buddies Activity

Weathering & Erosion Walk

Procedures

1. Take a walk with your Science Buddy around your neighborhood. Look for signs of weathering and erosion. *Weathering* is the breaking down of rocks into smaller pieces. *Erosion* is the movement of sand and soil by forces such as water or wind. You can find evidence of weathering and erosion by examining:
   - playgrounds
   - sidewalks
   - construction sites
   - driveways
   - gardens and flower beds
   - ditches
   - steep hillsides
   - stream beds

2. Choose three of your favorite examples and draw them on the Weathering and Erosion Log. Next to each illustration, write the location and a brief description of your observations.

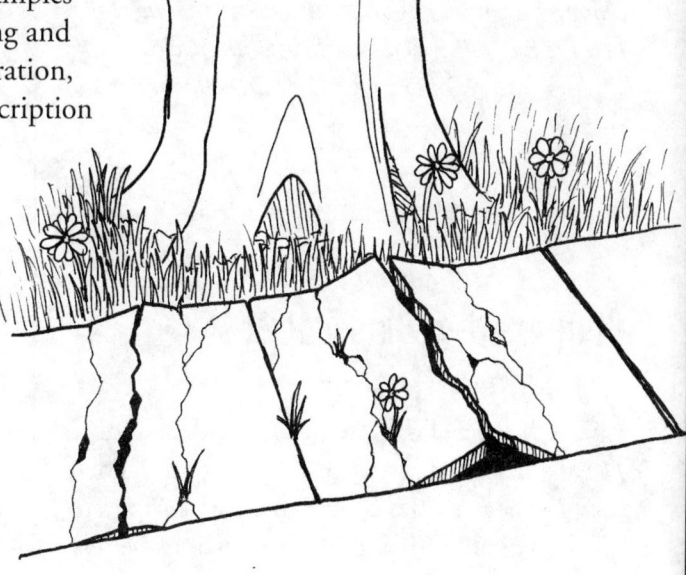

Due Date

Materials
- Weathering and Erosion Log
- Pencil
- Colored pencils, crayons, or markers
- Clipboard or notebook

Talk It Over

1. What is the difference between weathering and erosion?
2. Where does most erosion seem to occur?
3. What problems might be caused by weathering and erosion?
4. How can people can prevent erosion?

Science Buddies® Laura Candler • Kagan Publishing • 1-800-933-2667

Science Buddies Teacher's Page

Balloon Blast-Off

Overview

This activity is a simple demonstration of Newton's Third Law: For every action there is an equal and opposite reaction. When the air rushes out of the back of a balloon, the balloon zooms forward in the opposite direction. In this case, the air is the "action," and the "reaction" is the balloon's forward movement. This activity also illustrates the concept of jet propulsion. Rockets burn fuel very rapidly, and the pressure of the burning gas leaving rocket creates enough upward thrust to overcome the force of gravity.

Introducing the Activity

Before handing out the Science Buddies activity, if possible, show video footage of a Space Shuttle lift-off. Ask students to **Think-Pair-Share** the following question: *"How is the Space Shuttle able to overcome the force of gravity and lift off into space?"*

Do not give them the answer, but tell them that this Science Buddy activity may help them understand how rockets work.

Safety Notes: Discourage young children from blowing up balloons on their own. Balloons can become a choking hazard if the child accidentally inhales instead of exhales while attempting to inflate it.

Cooperative Classroom Follow Up

If time allows, demonstrate the balloon rocket activity in the classroom. Then have students turn to a partner for a **Pair Discussion** of the questions below.

Answers To Talk It Over

1. The air rushes out the end of the balloon towards the person holding it.
2. The balloon travels forward in the opposite direction from the person releasing it.
3. The action of the balloon is similar to that of a rocket because they both illustrate Newton's Third Law (for every action there is and equal and opposite reaction). The balloon's forward movement is caused by the air rushing out of its open end. In the same way, a rocket can lift off because of the force of the burning gases blasting from the rear of the spacecraft.

Science Buddies Teacher's Page

Science Buddies Activity

Balloon Blast-Off

Procedures

1. Cut a piece of string or thread about 15 feet long.
2. Tie one end to a door knob, the back of a chair, or any other stationary object.
3. Slip the other end of the string through the plastic straw.
4. Cut a 4-inch piece of tape and place it loosely over the straw. You will use this later.
5. Have your Science Buddy blow up the balloon and hold the end closed with your fingers. Do *not* tie a knot in the end.
6. Ask your Science Buddy to help you tape the balloon under the straw so that the open end is toward you.
7. Pull the straw and balloon to the beginning of the string and hold the string very tight.
8. What do you think will happen when you let go of the balloon? Try it and find out!

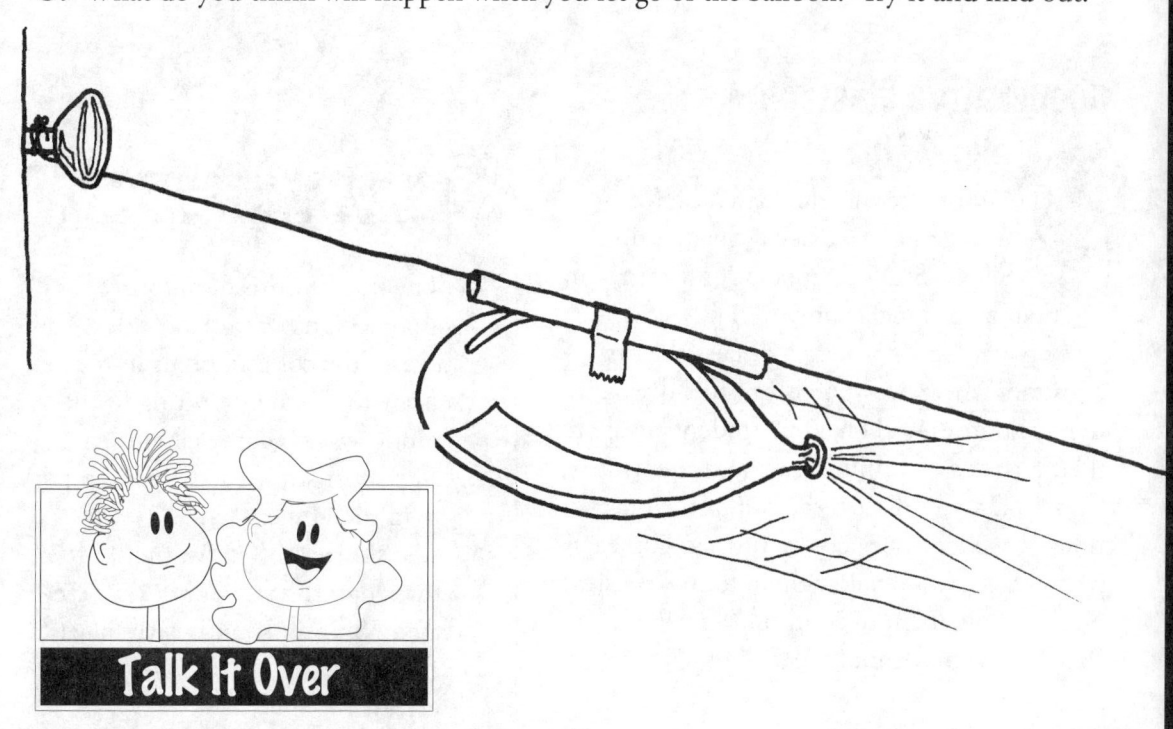

Due Date

Materials

- Thread or smooth string
- Plastic drinking straw
- Long balloon
- Tape

Talk It Over

1. In which direction did the air leave the balloon?
2. In which direction did the balloon travel?
3. How is the action of the balloon like that of a rocket?

Science Buddies® Laura Candler • Kagan Publishing • 1-800-933-2667

Science Buddies Teacher's Page

Straw Oboes

Overview

With "Straw Oboes" students explore sound with an ordinary straw. By shaping the end of a straw into a point and blowing into it, students create a simple woodwind instrument. Sound is created when air is made to vibrate. The tips of the straw and the column of air within the straw vibrate causing a humming sound.

Introducing the Activity

Before handing out the Science Buddies activity, ring a bell or tap a spoon against a glass. Ask students to **Think-Pair-Share** how the sound reaches their ears. Tell them that they will learn about sound by making a musical instrument out of a straw.

Safety Notes: Tell students not to become discouraged if it takes a little practice to make the straw oboe work. However, if they keep blowing into it without stopping to catch their breath, they may become dizzy. Tell them to stop periodically and breathe normally for a few minutes.

Cooperative Classroom Follow Up

On or after the due date, use **Mix-Freeze-Pair** to discuss the questions below. Students mix around the room, then on a signal from you they freeze. They pair with a partner to discuss the first question. After everyone is finished discussing their responses, call on a few students to share their answers. Continue having students mix, freeze, pair, and discuss the rest of the questions. To extend the activity further, give each person a straw and allow them to make a straw oboe. Allow them to continue **Mix-Freeze-Pair** as they play tunes for each other to guess.

Answers To Talk It Over

1. The pointed tips of the straw act as a reed and vibrate quickly. This causes the column of air in the straw to vibrate, resulting in the sound waves that reach our ears.
2. The short straw has a higher pitch than the longer straw.
3. The shorter the straw, the shorter the length of the sound wave created. Shorter sound waves have a higher pitch than longer sound waves.
4. The straw oboe is similar to instruments in the woodwind category, especially the clarinet or the oboe.

Science Buddies Activity

Straw Oboes

Procedures

1. Ask your Science Buddy to help you make a straw oboe. Flatten the end of one straw. Carefully snip the sides away, leaving a pointed tip as shown below.
2. Make a second straw oboe for your Science Buddy.
3. To make a sound, place the pointed end of the straw between your lips. Blow into the straw. If you don't hear a sound, try biting down gently on the straw. Keep trying different methods, but if you begin to feel dizzy stop blowing and breathe normally for a few minutes.
4. When you both can make a sound with your straws, trim the end of your own straw oboe so that it is only three inches long.
5. Do you think the short straw will sound any different from the long one? Try blowing into it to find out. Have your Science Buddy blow his or her straw oboe again to compare sounds.
6. Can you play a tune on your straw oboe? Play a simple song and see if your Science Buddy can guess the title. Let your Buddy play one for you to guess!

Due Date

Materials

- 2 drinking straws
- Scissors
- Ruler

Talk It Over

1. How do you think the straw oboe works?
2. How did the short straw oboe sound different from the long one?
3. Why do you think the two straws sounded different?
4. What musical instruments are similar to your straw oboe?

Science Buddies® Laura Candler • Kagan Publishing • 1-800-933-2667

Science Buddies Teacher's Page

Lever Logic

Overview

In this activity students explore first-class levers by constructing one from a ruler, two cups, and a pencil. By dropping pennies into a cup on one end, students determine the amount of force needed to lift a small object on the other end. Students experiment with various fulcrum positions by changing the position of the pencil and testing each location.

Introducing the Activity

Before handing out the Science Buddies activity, divide teams into two sets of partners for a **Pair Discussion**. Pose this question: *"If a child and an adult sit on opposite ends of a see-saw, is it possible for the child to lift the adult?"* In responding, students may mention the weights of the people, the position of each person on the see-saw, or the position of the see-saw on its crossbar.

Cooperative Classroom Follow Up

On or after the due date, use **Numbered Heads Together** to discuss the questions. Have students number off from 1-4 in their teams. Call out the first question and have students put their heads together in teams to discuss the answer. Allow several minutes of discussion, then randomly select a number and ask all students with that number to stand. Students may respond orally in turn or by writing answers on a team chalkboard. After a brief class discussion of the responses, repeat the steps with the remaining questions.

Answers To Talk It Over

1. The least amount of force is needed with the fulcrum placed at the 4 inch mark. As the fulcrum is moved closer to the load, *less* force is needed to lift that load. As the fulcrum is moved farther away, *greater* force is required. (If students moved the fulcrum closer than the 4 inch mark, they would have found even less force needed.)
2. Yes. No matter what the load, less force is needed to lift it when the fulcrum is placed close to the load.
3. Yes. If the heavy person scoots up closer to the crossbar and the light person remains at the end of the see-saw, the lighter person can lift the heavier one. Another solution is to physically move the see-saw's position on the crossbar so that one end is longer than the other one.
4. Examples of levers include can openers, car jacks, and hammers (when used to pull out nails).

66   Science Buddies® Laura Candler • Kagan Publishing • 1-800-933-2667

Science Buddies Activity

Lever Logic

Procedures

1. To make a lever, tape two identical small cups to the ends of a ruler. Lay the pencil on a flat surface and place the ruler across the top as shown below. The pencil will serve as the lever's *fulcrum*.
2. Put a small object (such as a salt shaker) in the cup taped over the 1 inch mark. This object will be the *load* you are lifting.
3. Adjust the ruler so that the fulcrum is positioned below the 4 inch mark. Predict how many pennies you will need to lift the load. The pennies will measure the force needed to lift the load.
4. Drop pennies in the force cup, one at a time, until the ruler balances. Record the fulcrum position (4 inches) and the number of pennies.
5. Repeat the activity *using the same load* but moving the fulcrum to 5 inches, 6 inches, 7 inches, and 8 inches. Make a chart of your results.

Due Date

Materials
- 1 ruler
- 1 pencil
- 2 identical small cups (paper or plastic)
- Small object
- 50 - 100 pennies
- Tape

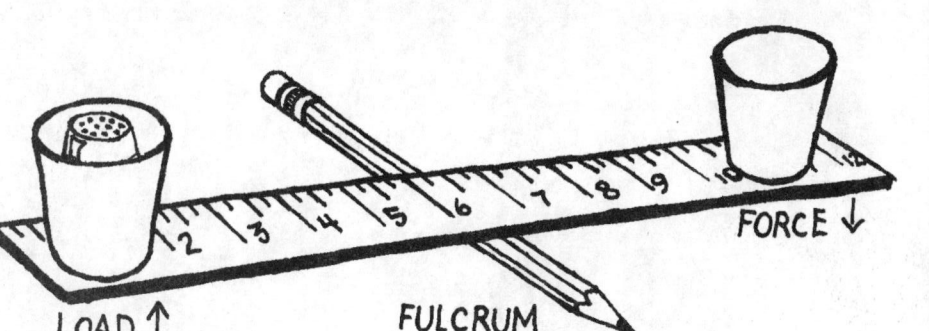

LOAD ↑   FULCRUM   FORCE ↓

Talk It Over

1. In which fulcrum position was the *least* amount of force needed to lift the load?
2. Would you get similar results with different loads? Try it and see!
3. Could a light person lift a heavier person on a see-saw? How?
4. What are some other examples of levers in everyday life?

Science Buddies® Laura Candler • Kagan Publishing • 1-800-933-2667

Chapter 4

Additional Reproducibles

This section contains patterns for several activities as well as helpful blackline masters for the teacher. You'll find a letter to parents, a record-keeping roster, Lab Report forms, Hints for Science Buddies, and more.

Science Buddies Program

Dear Parents,

We've been having fun with science at school. Now I'd like to invite you to share in some of that fun at home! We're starting a program called Science Buddies that offers you the chance to do just that.

At the beginning of each month your child will receive a worksheet describing a simple science activity to be completed at home with an adult "Science Buddy," such as a parent, grandparent, relative, or family friend. A student may keep the same Science Buddy all year or choose a new one each month.

Each worksheet lists the materials needed and describes how to perform the activity. The activities use only inexpensive and safe household items. Each activity ends with several "Talk It Over" questions for you to discuss with your child. We will also discuss the answers to these questions in class.

Please take a few minutes each month to share the fun of science discovery with your child. You'll soon be anticipating each month's activity with excitement!

Sincerely,

Hints for Science Buddies

Be enthusiastic about your science discoveries!

Check the Materials List and be sure you have all items. (Before buying any item, you might consider substituting a similar material.)

Ask your child to read the directions aloud as you work through the activity. It's a good idea to read through all the directions one time before starting the activity.

Be sure to complete the activity as a team, taking turns and sharing materials.

When making predictions before activities, allow your child to say his or her prediction first. Then give your own prediction. Reassure your child that it's okay to have different ideas about what might happen.

Don't worry if you don't know the answers to the discussion questions. Talk about possible answers with your child, asking them to state ideas as clearly as possible. Your child will discuss the questions in class on the due date, so be sure to ask them to share what was learned.

Encourage your child to extend the activities by doing similar experiments. Ask "I wonder what would happen if we . . ." Suggest that they share their findings with their classmates.

Be prepared to have fun messing around with science!

Science Buddies® Laura Candler • Kagan Publishing • 1-800-933-2667

Science Buddies Lab Report

Student _____

Science Buddy _____

Date _____

Activity Name _____

Predictions
Student _____
_____

Buddy _____
_____

What Happened _____
_____
_____
_____

What We Learned _____
_____
_____
_____

Other Ideas We Tried _____
_____
_____

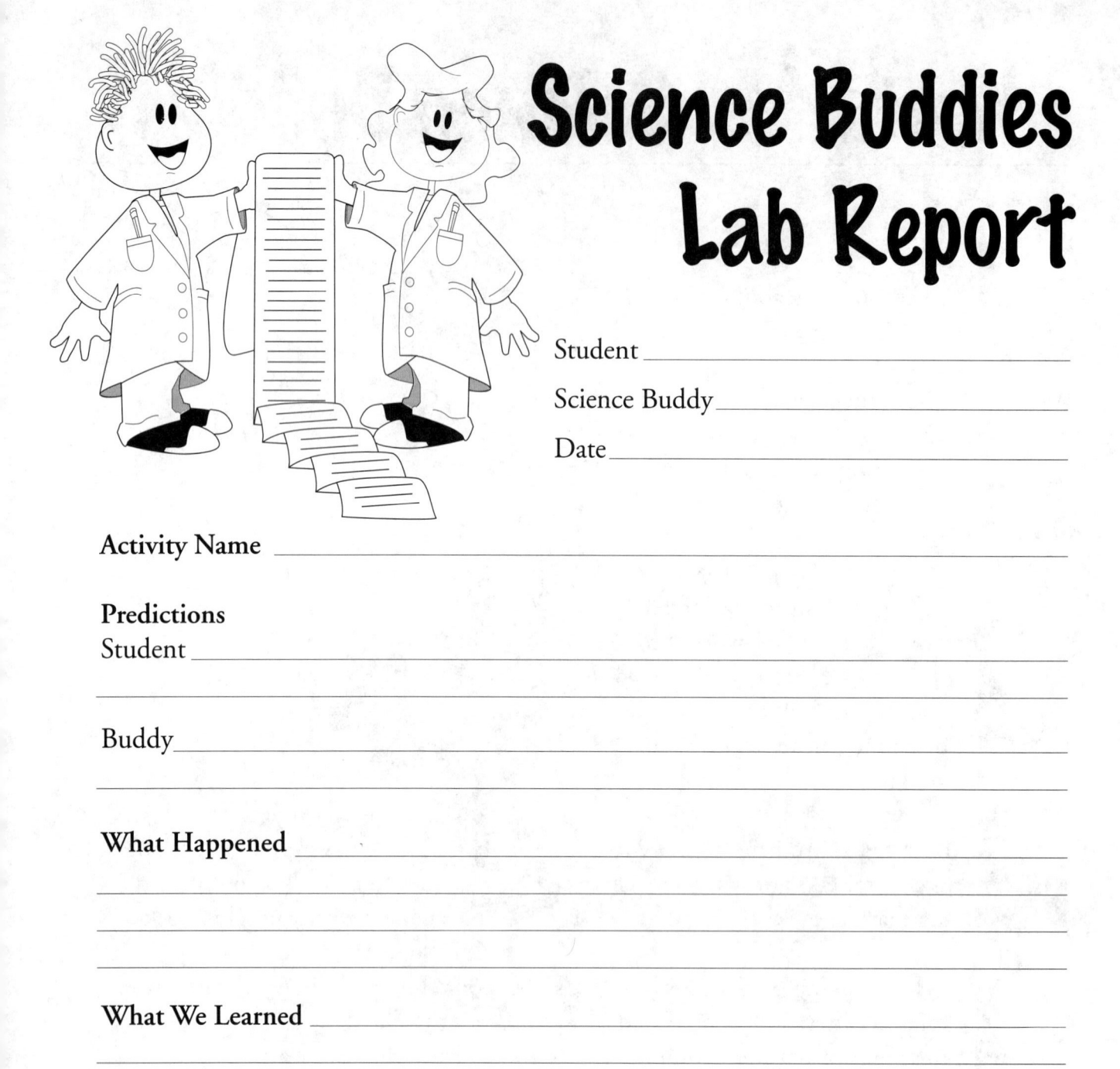

We completed this activity together.

_____   _____
      Student   Science Buddy

Science Buddies Lab Report

Date _____

Activity Name _____

Results _____
_____
_____
_____
_____
_____

We completed this activity together.

_____   _____
Student   Science Buddy

- -

Science Buddies Lab Report

Date _____

Activity Name _____

Results _____
_____
_____
_____
_____
_____

We completed this activity together.

_____   _____
Student   Science Buddy

Science Buddies® Laura Candler • Kagan Publishing • 1-800-933-2667

Science Buddies Roster

Names | Months

Super Science Buddy

has achieved the rank of Super Science Buddy for successfully completing all Science Buddies activities this school year.

Teacher

Date

Science Buddies® Laura Candler • Kagan Publishing • 1-800-933-2667

Color Spinner Patterns

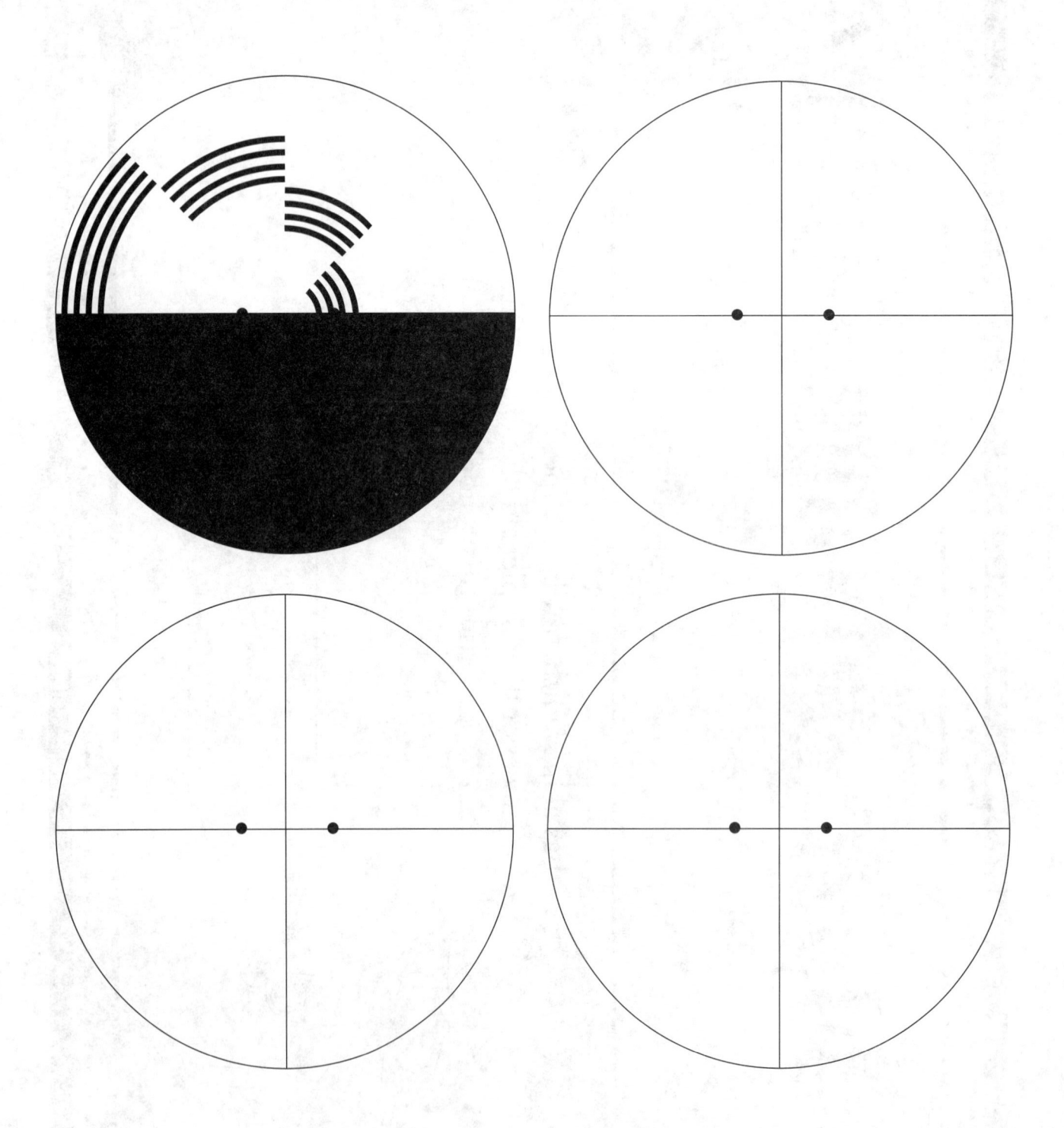

76

- **The Amazing Paper Kite**

- **The Amazing Paper Kite**

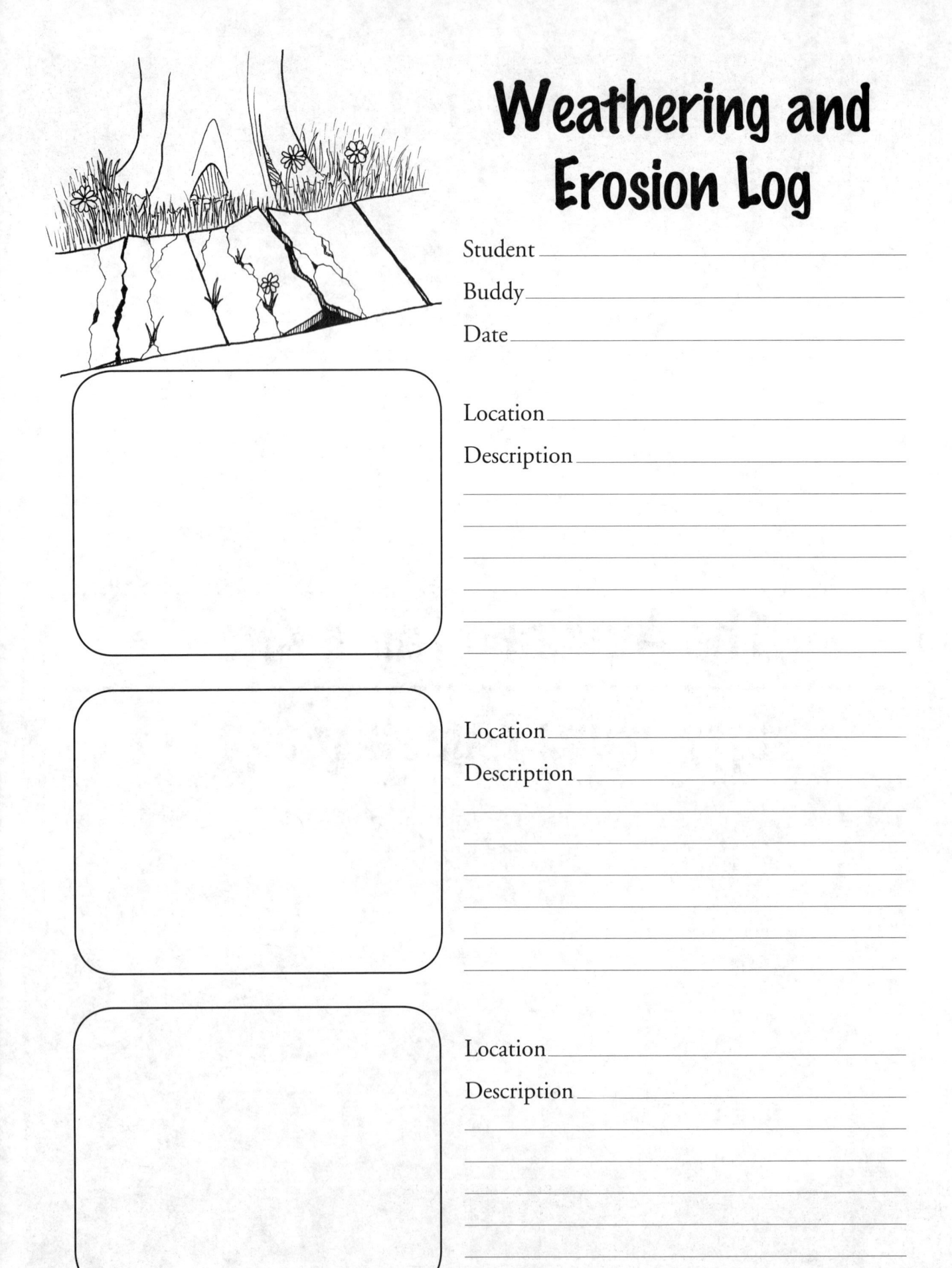

Weathering and Erosion Log

Student _____

Buddy _____

Date _____

Location _____

Description _____
_____
_____
_____
_____
_____
_____

Location _____

Description _____
_____
_____
_____
_____
_____
_____

Location _____

Description _____
_____
_____
_____
_____
_____
_____

Kagan

It's All About Engagement!

Kagan is the world leader in creating active engagement in the classroom. Learn how to engage your students and you will boost achievement, prevent discipline problems, and make learning more fun and meaningful. Come join Kagan for a workshop or call Kagan to **set up a workshop for your school or district.** Experience the power of a Kagan workshop. **Experience the engagement!**

SPECIALIZING IN:

★ **Cooperative Learning**
★ **Win-Win Discipline**
★ **Brain-Friendly Teaching**
★ **Multiple Intelligences**
★ **Thinking Skills**
★ **Kagan Coaching**

KAGAN PROFESSIONAL DEVELOPMENT

www.KaganOnline.com ★ 1(800) 266-7576

Kagan

It's All About Engagement!

**Kagan is your source
for active engagement in the classroom.**

Check out Kagan's line of books, SmartCards, software, electronics, and hands-on learning resources—all designed to boost engagement in your classroom.

Books

SmartCards

Spinners

Learning Chips

Posters

Learning Cubes

KAGAN PUBLISHING

www.KaganOnline.com ★ 1(800) 933-2667